Deblurring Images

Fundamentals of Algorithms

Editor-in-Chief: Nicholas J. Higham, University of Manchester

The SIAM series on Fundamentals of Algorithms is a collection of short user-oriented books on state-of-the-art numerical methods. Written by experts, the books provide readers with sufficient knowledge to choose an appropriate method for an application and to understand the method's strengths and limitations. The books cover a range of topics drawn from numerical analysis and scientific computing. The intended audiences are researchers and practitioners using the methods and upper level undergraduates in mathematics, engineering, and computational science.

Books in this series not only provide the mathematical background for a method or class of methods used in solving a specific problem but also explain how the method can be developed into an algorithm and translated into software. The books describe the range of applicability of a method and give guidance on troubleshooting solvers and interpreting results. The theory is presented at a level accessible to the practitioner. MATLAB® software is the preferred language for codes presented since it can be used across a wide variety of platforms and is an excellent environment for prototyping, testing, and problem solving.

The series is intended to provide guides to numerical algorithms that are readily accessible, contain practical advice not easily found elsewhere, and include understandable codes that implement the algorithms.

Series Volumes

Chan, R. H.-F. and Jin, X.-Q., *An Introduction to Iterative Toeplitz Solvers*
Eldén, L., *Matrix Methods in Data Mining and Pattern Recognition*
Hansen, P. C., Nagy, J. G., and O'Leary, D. P., *Deblurring Images: Matrices, Spectra, and Filtering*
Davis, T. A., *Direct Methods for Sparse Linear Systems*
Kelley, C. T., *Solving Nonlinear Equations with Newton's Method*

Per Christian Hansen
Technical University of Denmark
Lyngby, Denmark

James G. Nagy
Emory University
Atlanta, Georgia

Dianne P. O'Leary
University of Maryland
College Park, Maryland

Deblurring Images

Matrices, Spectra, and Filtering

Society for Industrial and Applied Mathematics
Philadelphia

10 9 8 7 6 5 4 3 2

Library of Congress Cataloging-in-Publication Data

Hansen, Per Christian.
Deblurring images : matrices, spectra, and filtering / Per Christian Hansen, James G. Nagy, Dianne P. O'Leary.
p. cm. -- (Fundamentals of algorithms)
Includes bibliographical references and index.
ISBN 978-0-898716-18-4 (pbk.)
1. Image processing--Mathematical models. I. Title.

TA1637.H364 2006
621.36'7015118--dc22

2006050450

To our teachers and our students

Contents

Preface ix

How to Get the Software xii

List of Symbols xiii

1 The Image Deblurring Problem 1
- 1.1 How Images Become Arrays of Numbers . . . 2
- 1.2 A Blurred Picture and a Simple Linear Model . . . 4
- 1.3 A First Attempt at Deblurring . . . 5
- 1.4 Deblurring Using a General Linear Model . . . 7

2 Manipulating Images in MATLAB 13
- 2.1 Image Basics . . . 13
- 2.2 Reading, Displaying, and Writing Images . . . 14
- 2.3 Performing Arithmetic on Images . . . 16
- 2.4 Displaying and Writing Revisited . . . 18

3 The Blurring Function 21
- 3.1 Taking Bad Pictures . . . 21
- 3.2 The Matrix in the Mathematical Model . . . 22
- 3.3 Obtaining the PSF . . . 24
- 3.4 Noise . . . 28
- 3.5 Boundary Conditions . . . 29

4 Structured Matrix Computations 33
- 4.1 Basic Structures . . . 34
  - 4.1.1 One-Dimensional Problems . . . 34
  - 4.1.2 Two-Dimensional Problems . . . 37
  - 4.1.3 Separable Two-Dimensional Blurs . . . 38
- 4.2 BCCB Matrices . . . 40
  - 4.2.1 Spectral Decomposition of a BCCB Matrix . . . 41
  - 4.2.2 Computations with BCCB Matrices . . . 43
- 4.3 BTTB + BTHB + BHTB + BHHB Matrices . . . 44
- 4.4 Kronecker Product Matrices . . . 48

4.4.1 Constructing the Kronecker Product from the PSF 48
4.4.2 Matrix Computations with Kronecker Products 49
4.5 Summary of Fast Algorithms 51
4.6 Creating Realistic Test Data 52

5 SVD and Spectral Analysis 55
5.1 Introduction to Spectral Filtering 55
5.2 Incorporating Boundary Conditions 57
5.3 SVD Analysis 58
5.4 The SVD Basis for Image Reconstruction 61
5.5 The DFT and DCT Bases 63
5.6 The Discrete Picard Condition 67

6 Regularization by Spectral Filtering 71
6.1 Two Important Methods 71
6.2 Implementation of Filtering Methods 74
6.3 Regularization Errors and Perturbation Errors 77
6.4 Parameter Choice Methods 79
6.5 Implementation of GCV 82
6.6 Estimating Noise Levels 84

7 Color Images, Smoothing Norms, and Other Topics 87
7.1 A Blurring Model for Color Images 87
7.2 Tikhonov Regularization Revisited 90
7.3 Working with Partial Derivatives 92
7.4 Working with Other Smoothing Norms 96
7.5 Total Variation Deblurring 97
7.6 Blind Deconvolution 99
7.7 When Spectral Methods Cannot Be Applied 100

Appendix: MATLAB Functions 103
1. TSVD Regularization Methods 103
Periodic Boundary Conditions 103
Reflexive Boundary Conditions 104
Separable Two-Dimensional Blur 106
Choosing Regularization Parameters 107
2. Tikhonov Regularization Methods 108
Periodic Boundary Conditions 108
Reflexive Boundary Conditions 109
Separable Two-Dimensional Blur 111
Choosing Regularization Parameters 112
3. Auxiliary Functions 113

Bibliography 121

Index 127

Preface

There is nothing worse than a sharp image of a fuzzy concept.
– Ansel Adams

Whoever controls the media—the images—controls the culture.
– Allen Ginsberg

This book is concerned with deconvolution methods for image deblurring, that is, computational techniques for reconstruction of blurred images based on a concise mathematical model for the blurring process. The book describes the algorithms and techniques collectively known as spectral filtering methods, in which the singular value decomposition—or a similar decomposition with spectral properties—is used to introduce the necessary regularization or filtering in the reconstructed image.

The main purpose of the book is to give students and engineers an understanding of the linear algebra behind the filtering methods. Readers in applied mathematics, numerical analysis, and computational science will be exposed to modern techniques to solve realistic large-scale problems in image deblurring.

The book is intended for beginners in the field of image restoration and regularization. While the underlying mathematical model is an ill-posed problem in the form of an integral equation of the first kind (for which there is a rich theory), we have chosen to keep our formulations in terms of matrices, vectors, and matrix computations. Our reasons for this choice of formulation are twofold: (1) the linear algebra terminology is more accessible to many of our readers, and (2) it is much closer to the computational tools that are used to solve the given problems. Throughout the book we give references to the literature for more details about the problems, the techniques, and the algorithms—including the insight that is obtained from studying the underlying ill-posed problems.

All the methods presented in this book belong to the general class of regularization methods, which are methods specially designed for solving ill-posed problems. We do not require the reader to be familiar with these regularization methods or with ill-posed problems. For readers who already have this knowledge, we aim to give a new and practical perspective on the issues of using regularization methods to solve real problems.

We will assume that the reader is familiar with MATLAB and also, preferably, has access to the MATLAB Image Processing Toolbox (IPT). The topics covered in our book are well suited for computer demonstrations, and our aim is that the reader will be able to start deblurring images while reading the book. MATLAB provides a convenient and widespread computational platform for doing numerical computations, and therefore it is natural to use it for the examples and algorithms presented here. Without too much pain, a user can then make more dedicated and efficient computer implementations if there is a need for it, based on the MATLAB "templates" presented in this book.

We will also assume that the reader is familiar with basic concepts of linear algebra and matrix computations, including the singular value decomposition and orthogonal transformations. We do not require the signal processing background that is often needed in classical books on image processing.

The book starts with a short chapter that introduces the fundamental problem of image deblurring and the spectral filtering methods for computing reconstructions. The chapter sets up the basic notation for the linear system of equations associated with the blurring model, and also introduces the most important tools, techniques, and concepts needed for the remaining chapters.

Chapter 2 explains how to work with images of various formats in MATLAB. We explain how to load and store the images, and how to perform mathematical operations on them.

Chapter 3 gives a description of the image blurring process. We derive the mathematical model for the point spread function (PSF) that describes the blurring due to different sources, and we discuss some topics related to the boundary conditions that must always be specified.

Chapter 4 gives a thorough description of structured matrix computations. We introduce circulant, Toeplitz, and Hankel matrices, as well as Kronecker products. We show how these structures reflect the PSF, and how operations with these matrices can be performed efficiently by means of the FFT algorithm.

Chapter 5 builds up an understanding of the mechanisms and difficulties associated with image deblurring, expressed in terms of spectral decompositions, thus setting the stage for the reconstruction algorithms.

Chapter 6 explains how regularization, in the form of spectral filtering, is applied to the image deblurring problem. In addition to covering several spectral filtering methods and their implementations, we also discuss methods for choosing the regularization parameter that controls the smoothing.

Finally, Chapter 7 gives an introduction to other aspects of deblurring methods and techniques that we cannot cover in depth in this book.

Throughout the book we have included Very Important Points (VIPs) to summarize the presentation and Pointers to provide additional information and references. We also provide Challenges so that the reader can gain experience with the methods we discuss. We hope that readers have fun with these, especially in deblurring the mystery image of Challenge 2.

The images and MATLAB functions discussed in the book, as well as additional Challenges and other material, can be found at

`www.siam.org/books/fa03`

We are most grateful for the help and support we have received in writing this book. The U.S. National Science Foundation and the Danish Research Agency provided funding for much of the work upon which this book is based. Linda Thiel, Sara Murphy, and others on the SIAM staff patiently worked with us in preparing the manuscript for print. The referees and other readers provided many helpful comments. We would like to acknowledge, in particular, Julianne Chung, Martin Hanke-Bourgeois, Nicholas Higham, Stephen Marsland, Robert Plemmons, and Zdeněk Strakoš. Nicola Mastronardi graciously invited us to present a course based on this book at the Third International School in Numerical Linear Algebra and Applications, Monopoli, Italy, September 2005, and the book benefited from the suggestions of the participants and the experience gained there. Thank you to all.

Per Christian Hansen
James G. Nagy
Dianne P. O'Leary

Lyngby, Atlanta, and College Park, 2006

How to Get the Software

This book is accompanied by a small package of MATLAB software as well as some test images. The software and images are available from SIAM at the URL

`www.siam.org/books/fa03`

The material on the website is organized as follows:

- HNO FUNCTIONS: a small MATLAB package, written by us, which implements all the image deblurring algorithms presented in the book. It requires MATLAB version 6.5 or newer versions. The package also includes several auxiliary functions, e.g., for creating point spread functions.
- CHALLENGE FILES: the files for the Challenges in the book, designed to let the reader experiment with the methods.
- ADDITIONAL IMAGES: a small collection with some additional images which can be used for tests and experiments.
- ADDITIONAL MATERIAL: background material about matrix decompositions used in this book.
- ADDITIONAL CHALLENGES: a small collection of additional Challenges related to the book.

We invite readers to contribute additional challenges and images.

MATLAB can be obtained from

The MathWorks, Inc.
3 Apple Hill Drive
Natick, MA 01760-2098
(508) 647-7000
Fax: (508) 647-7001
Email: `info@mathworks.com`
URL: `http://www.mathworks.com`

List of Symbols

We begin with a survey of the notation and image deblurring jargon used in this book. Throughout, an image (grayscale or color) is always referred to as an *image array*, having in mind its natural representation in MATLAB. For the same reason we use the term *PSF array* for the image of the point spread function. The phrase *matrix* is reserved for use in connection with the linear systems of equations that form the basis for our methods. The *fast transforms* (FFT and DCT) used in our algorithms are always computed by means of efficient implementations although, for notational reasons, we often represent them by matrices.

All of the main symbols used in the book are listed here. Capital boldface always denotes a matrix or an array, while small boldface denotes a vector and a plain italic typeface denotes a scalar or an integer.

IMAGE SYMBOLS	
Image array (always $m \times n$)	$\mathbf{B}$, $\mathbf{X}$
Noise "image" (always $m \times n$)	$\mathbf{E}$
PSF array (always $m \times n$)	$\mathbf{P}$
Dimensions of image array	$m \times n$
LINEAR ALGEBRA SYMBOLS	
Matrix (always $N \times N$)	$\mathbf{A}$
Matrix dimension	$N = m \cdot n$
Matrix element (i, j) of $\mathbf{A}$	a_{ij}
Column i of matrix $\mathbf{A}$	$\mathbf{a}_i$
Identity matrix (order ℓ)	$\mathbf{I}_\ell$
Vector	$\mathbf{b}$, $\mathbf{e}$, $\mathbf{x}$
Vector element i of $\mathbf{x}$	x_i
Standard unit vector (ith column of identity matrix)	$\mathbf{e}_i$
2-norm, p-norm, Frobenius norm	$\|\cdot\|_2$, $\|\cdot\|_p$, $\|\cdot\|_F$

Special Matrices	
Boundary conditions matrix	$\mathbf{A}_{\mathrm{BC}}$
Discrete derivative matrix	$\mathbf{D}$
Matrix for zero boundary conditions	$\mathbf{A}_0$
Color blurring matrix (always 3×3)	$\mathbf{A}_{\mathrm{color}}$
Column blurring matrix	$\mathbf{A}_{\mathrm{c}}$
Row blurring matrix	$\mathbf{A}_{\mathrm{r}}$
Shift matrices	$\mathbf{Z}_1, \mathbf{Z}_2$

Spectral Decomposition	
Matrix of eigenvectors	$\widetilde{\mathbf{U}}$
Diagonal matrix of eigenvalues	$\mathbf{\Lambda}$

Singular Value Decomposition	
Matrix of left singular vectors	$\mathbf{U}$
Matrix of right singular vectors	$\mathbf{V}$
Diagonal matrix of singular values	$\mathbf{\Sigma}$
Left singular vector	$\mathbf{u}_i$
Right singular vector	$\mathbf{v}_i$
Singular value	σ_i

Regularization	
Filter factor	ϕ_i
Diagonal matrix of filter factors	$\mathbf{\Phi} = \mathrm{diag}(\phi_i)$
Truncation parameter for TSVD	k
Regularization parameter for Tikhonov	α

Other Symbols	
Kronecker product	$\otimes$
Stacking columns: vec notation	$\mathrm{vec}(\cdot)$
Complex conjugation	$\mathrm{conj}(\cdot)$
Discrete cosine transform (DCT) matrix (two-dimensional)	$\mathbf{C} = \mathbf{C}_{\mathrm{r}} \otimes \mathbf{C}_{\mathrm{c}}$
Discrete Fourier transform (DFT) matrix (two-dimensional)	$\mathbf{F} = \mathbf{F}_{\mathrm{r}} \otimes \mathbf{F}_{\mathrm{c}}$

Chapter 1

The Image Deblurring Problem

You cannot depend on your eyes when your imagination is out of focus.
– Mark Twain

When we use a camera, we want the recorded image to be a faithful representation of the scene that we see—but every image is more or less blurry. Thus, image deblurring is fundamental in making pictures sharp and useful.

A digital image is composed of picture elements called pixels. Each pixel is assigned an intensity, meant to characterize the color of a small rectangular segment of the scene. A small image typically has around $256^2 = 65536$ pixels while a high-resolution image often has 5 to 10 million pixels. Some blurring always arises in the recording of a digital image, because it is unavoidable that scene information "spills over" to neighboring pixels. For example, the optical system in a camera lens may be out of focus, so that the incoming light is smeared out. The same problem arises, for example, in astronomical imaging where the incoming light in the telescope has been slightly bent by turbulence in the atmosphere. In these and similar situations, the inevitable result is that we record a blurred image.

In image deblurring, we seek to recover the original, sharp image by using a mathematical model of the blurring process. The key issue is that some information on the lost details is indeed present in the blurred image—but this information is "hidden" and can only be recovered if we know the details of the blurring process.

Unfortunately there is no hope that we can recover the original image exactly! This is due to various unavoidable errors in the recorded image. The most important errors are fluctuations in the recording process and approximation errors when representing the image with a limited number of digits. The influence of this noise puts a limit on the size of the details that we can hope to recover in the reconstructed image, and the limit depends on both the noise and the blurring process.

POINTER. Image enhancement is used in the restoration of older movies. For example, the original Star Wars trilogy was enhanced for release on DVD. These methods are not model based and therefore not covered in this book. See [33] for more information.

POINTER. Throughout the book, we provide example images and MATLAB code. This material can be found on the book's website:

`www.siam.org/books/fa03`

For readers needing an introduction to MATLAB programming, we suggest the excellent book by Higham and Higham [27].

One of the challenges of image deblurring is to devise efficient and reliable algorithms for recovering as much information as possible from the given data. This chapter provides a brief introduction to the basic image deblurring problem and explains why it is difficult. In the following chapters we give more details about techniques and algorithms for image deblurring.

MATLAB is an excellent environment in which to develop and experiment with filtering methods for image deblurring. The basic MATLAB package contains many functions and tools for this purpose, but in some cases it is more convenient to use routines that are only available from the *Signal Processing Toolbox* (SPT) and the *Image Processing Toolbox* (IPT). We will therefore use these toolboxes when convenient. When possible, we provide alternative approaches that require only core MATLAB commands in case the reader does not have access to the toolboxes.

1.1 How Images Become Arrays of Numbers

Having a way to represent images as arrays of numbers is crucial to processing images using mathematical techniques. Consider the following 9×16 array:

$$\begin{bmatrix} 0&0&0&0&0&0&0&0&0&0&0&0&0&0&0&0 \\ 0&8&8&0&0&0&0&4&4&0&0&0&0&0&0&0 \\ 0&8&8&0&0&0&0&4&4&0&3&3&3&3&3&0 \\ 0&8&8&0&0&0&0&4&4&0&3&3&3&3&3&0 \\ 0&8&8&0&0&0&0&4&4&0&3&3&3&3&3&0 \\ 0&8&8&0&0&0&0&4&4&0&3&3&3&3&3&0 \\ 0&8&8&8&8&8&0&4&4&0&3&3&3&3&3&0 \\ 0&8&8&8&8&8&0&4&4&0&0&0&0&0&0&0 \\ 0&0&0&0&0&0&0&0&0&0&0&0&0&0&0&0 \end{bmatrix}.$$

If we enter this into a MATLAB variable `X` and display the array with the commands `imagesc(X), axis image, colormap(gray)`, then we obtain the picture shown at the left of Figure 1.1. The entries with value 8 are displayed as white, entries equal to zero are black, and values in between are shades of gray.

Color images can be represented using various formats; the RGB format stores images as three components, which represent their intensities on the red, green, and blue scales. A pure red color is represented by the intensity values $(1, 0, 0)$ while, for example, the values $(1, 1, 0)$ represent yellow and $(0, 0, 1)$ represent blue; other colors can be obtained with different choices of intensities. Hence, to represent a color image, we need three values per pixel. For example, if `X` is a multidimensional MATLAB array of dimensions $9 \times 16 \times 3$

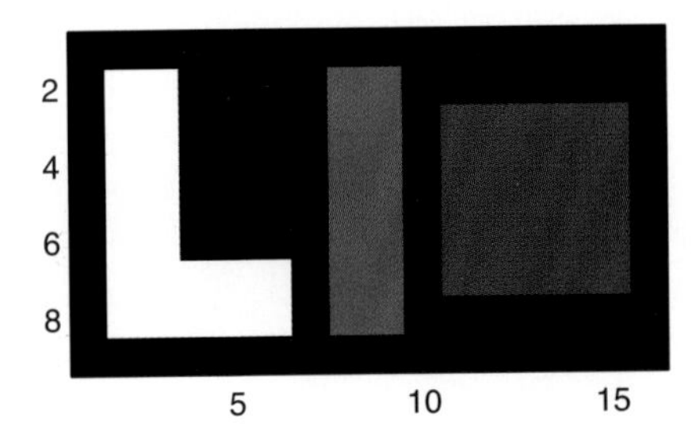

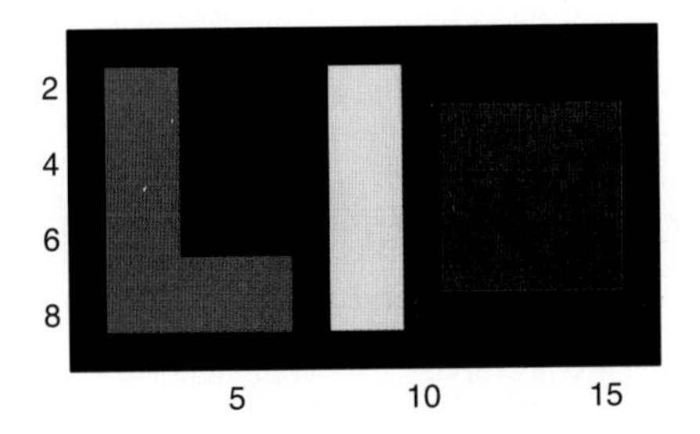

Figure 1.1. *Images created by displaying arrays of numbers.*

defined as

$$
\texttt{X(:,:,1)} = \begin{bmatrix}
0&0&0&0&0&0&0&0&0&0&0&0&0&0&0&0\\
0&1&1&0&0&0&0&1&1&0&0&0&0&0&0&0\\
0&1&1&0&0&0&0&1&1&0&0&0&0&0&0&0\\
0&1&1&0&0&0&0&1&1&0&0&0&0&0&0&0\\
0&1&1&0&0&0&0&1&1&0&0&0&0&0&0&0\\
0&1&1&0&0&0&0&1&1&0&0&0&0&0&0&0\\
0&1&1&1&1&1&0&1&1&0&0&0&0&0&0&0\\
0&1&1&1&1&1&0&1&1&0&0&0&0&0&0&0\\
0&0&0&0&0&0&0&0&0&0&0&0&0&0&0&0
\end{bmatrix},
$$

$$
\texttt{X(:,:,2)} = \begin{bmatrix}
0&0&0&0&0&0&0&0&0&0&0&0&0&0&0&0\\
0&0&0&0&0&0&0&1&1&0&0&0&0&0&0&0\\
0&0&0&0&0&0&0&1&1&0&0&0&0&0&0&0\\
0&0&0&0&0&0&0&1&1&0&0&0&0&0&0&0\\
0&0&0&0&0&0&0&1&1&0&0&0&0&0&0&0\\
0&0&0&0&0&0&0&1&1&0&0&0&0&0&0&0\\
0&0&0&0&0&0&0&1&1&0&0&0&0&0&0&0\\
0&0&0&0&0&0&0&1&1&0&0&0&0&0&0&0\\
0&0&0&0&0&0&0&0&0&0&0&0&0&0&0&0
\end{bmatrix},
$$

$$
\texttt{X(:,:,3)} = \begin{bmatrix}
0&0&0&0&0&0&0&0&0&0&0&0&0&0&0&0\\
0&0&0&0&0&0&0&0&0&0&0&0&0&0&0&0\\
0&0&0&0&0&0&0&0&0&0&1&1&1&1&1&0\\
0&0&0&0&0&0&0&0&0&0&1&1&1&1&1&0\\
0&0&0&0&0&0&0&0&0&0&1&1&1&1&1&0\\
0&0&0&0&0&0&0&0&0&0&1&1&1&1&1&0\\
0&0&0&0&0&0&0&0&0&0&1&1&1&1&1&0\\
0&0&0&0&0&0&0&0&0&0&0&0&0&0&0&0\\
0&0&0&0&0&0&0&0&0&0&0&0&0&0&0&0
\end{bmatrix},
$$

then we can display this image, in color, with the command `imagesc(X)`, obtaining the second picture shown in Figure 1.1. This brings us to our first Very Important Point (VIP).

> **VIP 1.** A digital image is a two- or three-dimensional array of numbers representing intensities on a grayscale or color scale.

Most of this book is concerned with grayscale images. However, the techniques carry over to color images, and in Chapter 7 we extend our notation and models to color images.

1.2 A Blurred Picture and a Simple Linear Model

Before we can deblur an image, we must have a mathematical model that relates the given blurred image to the unknown true image. Consider the example shown in Figure 1.2. The left is the "true" scene, and the right is a blurred version of the same image. The blurred image is precisely what would be recorded in the camera if the photographer forgot to focus the lens.

Figure 1.2. *A sharp image (left) and the corresponding blurred image (right).*

Grayscale images, such as the ones in Figure 1.2, are typically recorded by means of a CCD (charge-coupled device), which is an array of tiny detectors, arranged in a rectangular grid, able to record the amount, or intensity, of the light that hits each detector. Thus, as explained above, we can think of a grayscale digital image as a rectangular $m \times n$ array, whose entries represent light intensities captured by the detectors. To fix notation,

$\mathbf{X} \in \mathbb{R}^{m\times n}$ represents the desired sharp image, while

$\mathbf{B} \in \mathbb{R}^{m\times n}$ denotes the recorded blurred image.

Let us first consider a simple case where the blurring of the columns in the image is independent of the blurring of the rows. When this is the case, then there exist two matrices $\mathbf{A}_\mathrm{c} \in \mathbb{R}^{m\times m}$ and $\mathbf{A}_\mathrm{r} \in \mathbb{R}^{n\times n}$, such that we can express the relation between the sharp and blurred images as

$$\mathbf{A}_\mathrm{c}\,\mathbf{X}\,\mathbf{A}_\mathrm{r}^T = \mathbf{B}. \tag{1.1}$$

The left multiplication with the matrix $\mathbf{A}_\mathrm{c}$ applies the same *vertical* blurring operation to all n columns $\mathbf{x}_j$ of $\mathbf{X}$, because

$$\mathbf{A}_\mathrm{c}\,\mathbf{X} = \mathbf{A}_\mathrm{c}\begin{bmatrix} \mathbf{x}_1 & \mathbf{x}_2 & \cdots & \mathbf{x}_n \end{bmatrix} = \begin{bmatrix} \mathbf{A}_\mathrm{c}\mathbf{x}_1 & \mathbf{A}_\mathrm{c}\mathbf{x}_2 & \cdots & \mathbf{A}_\mathrm{c}\mathbf{x}_n \end{bmatrix}.$$

Similarly, the right multiplication with $\mathbf{A}_\mathrm{r}^T$ applies the same *horizontal* blurring to all m rows of $\mathbf{X}$. Since matrix multiplication is associative, i.e., $(\mathbf{A}_\mathrm{c}\,\mathbf{X})\,\mathbf{A}_\mathrm{r}^T = \mathbf{A}_\mathrm{c}\,(\mathbf{X}\,\mathbf{A}_\mathrm{r}^T)$, it does

POINTER. Image deblurring is much more than just a useful tool for our vacation pictures. For example, analysis of astronomical images gives clues to the behavior of the universe. At a more mundane level, barcode readers used in supermarkets and by shipping companies must be able to compensate for imperfections in the scanner optics; see Wittman [63] for more information.

not matter in which order we perform the two blurring operations. The reason for our use of the transpose of the matrix $\mathbf{A}_\mathrm{r}$ will be clear later, when we return to this blurring model and matrix formulations.

1.3 A First Attempt at Deblurring

If the image blurring model is of the simple form $\mathbf{A}_\mathrm{c}\,\mathbf{X}\,\mathbf{A}_\mathrm{r}^T = \mathbf{B}$, then one might think that the naïve solution

$$\mathbf{X}_{\text{naïve}} = \mathbf{A}_\mathrm{c}^{-1}\mathbf{B}\,\mathbf{A}_\mathrm{r}^{-T}$$

will yield the desired reconstruction, where $\mathbf{A}_\mathrm{r}^{-T} = (\mathbf{A}_\mathrm{r}^T)^{-1} = (\mathbf{A}_\mathrm{r}^{-1})^T$. Figure 1.3 illustrates that this is probably not such a good idea; the reconstructed image does not appear to have any features of the true image!

Figure 1.3. *The naïve reconstruction of the pumpkin image in Figure* 1.2, *obtained by computing* $\mathbf{X}_{\textit{naïve}} = \mathbf{A}_\mathrm{c}^{-1}\mathbf{B}\,\mathbf{A}_\mathrm{r}^{-T}$ *via Gaussian elimination on both* $\mathbf{A}_\mathrm{c}$ *and* $\mathbf{A}_\mathrm{r}$. *Both matrices are ill-conditioned, and the image* $\mathbf{X}_{\textit{naïve}}$ *is dominated by the influence from rounding errors as well as errors in the blurred image* **B**.

To understand why this naïve approach fails, we must realize that the blurring model in (1.1) is not quite correct, because we have ignored several types of errors.

Let us take a closer look at what is represented by the image **B**. First, let $\mathbf{B}_{\text{exact}} = \mathbf{A}_\mathrm{c}\,\mathbf{X}\,\mathbf{A}_\mathrm{r}^T$ represent the ideal blurred image, ignoring all kinds of errors. Because the blurred image is collected by a mechanical device, it is inevitable that small random errors (noise) will be present in the recorded data. Moreover, when the image is digitized, it is represented by a finite (and typically small) number of digits. Thus the recorded blurred image **B** is really given by

$$\mathbf{B} = \mathbf{B}_{\text{exact}} + \mathbf{E} = \mathbf{A}_\mathrm{c}\,\mathbf{X}\,\mathbf{A}_\mathrm{r}^T + \mathbf{E}, \tag{1.2}$$

where the noise image $\mathbf{E}$ (of the same dimensions as $\mathbf{B}$) represents the noise and the quantization errors in the recorded image. Consequently the naïve reconstruction is given by

$$\mathbf{X}_{\text{naïve}} = \mathbf{A}_{\text{c}}^{-1}\mathbf{B}\,\mathbf{A}_{\text{r}}^{-T} = \mathbf{A}_{\text{c}}^{-1}\mathbf{B}_{\text{exact}}\,\mathbf{A}_{\text{r}}^{-T} + \mathbf{A}_{\text{c}}^{-1}\mathbf{E}\,\mathbf{A}_{\text{r}}^{-T}$$

and therefore

$$\mathbf{X}_{\text{naïve}} = \mathbf{X} + \mathbf{A}_{\text{c}}^{-1}\mathbf{E}\,\mathbf{A}_{\text{r}}^{-T}, \tag{1.3}$$

where the term $\mathbf{A}_{\text{c}}^{-1}\mathbf{E}\,\mathbf{A}_{\text{r}}^{-T}$, which we can informally call inverted noise, represents the contribution to the reconstruction from the additive noise. This inverted noise will dominate the solution if the second term $\mathbf{A}_{\text{c}}^{-1}\mathbf{E}\,\mathbf{A}_{\text{r}}^{-T}$ in (1.3) has larger elements than the first term $\mathbf{X}$. Unfortunately, in many situations, as in Figure 1.3, the inverted noise indeed dominates.

Apparently, image deblurring is not as simple as it first appears. We can now state the purpose of our book more precisely, namely, to describe effective deblurring methods that are able to handle correctly the inverted noise.

CHALLENGE 1.

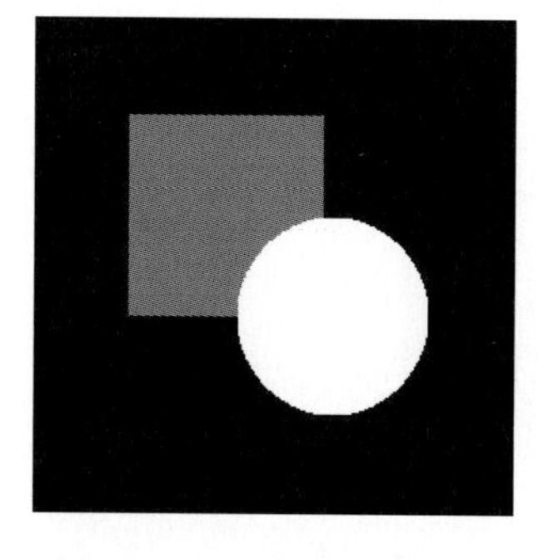

The exact and blurred images `X` and `B` in the above figure can be constructed in MATLAB by calling

```
[B, Ac, Ar, X] = challenge1(m, n, noise);
```

with `m = n = 256` and `noise = 0.01`. Try to deblur the image `B` using

```
Xnaive = Ac\B / Ar';
```

To display a grayscale image, say, `X`, use the commands

```
imagesc(X), axis image, colormap gray
```

How large can you choose the parameter `noise` before the inverted noise dominates the deblurred image? Does this value of `noise` depend on the image size?

CHALLENGE 2.

The above image B as well as the blurring matrices `Ac` and `Ar` are given in the file `challenge2.mat`. Can you deblur this image with the naïve approach, so that you can read the text in it?

As you learn more throughout the book, use Challenges 1 and 2 as examples to test your skills and learn more about the presented methods.

CHALLENGE 3. For the simple model $\mathbf{B} = \mathbf{A}_c \mathbf{X} \mathbf{A}_r^T + \mathbf{E}$ it is easy to show that the relative error in the naïve reconstruction $\mathbf{X}_{\text{naïve}} = \mathbf{A}_c^{-1}\mathbf{B}\mathbf{A}_r^{-T}$ satisfies

$$\frac{\|\mathbf{X}_{\text{naïve}} - \mathbf{X}\|_F}{\|\mathbf{X}\|_F} \leq \text{cond}(\mathbf{A}_c)\,\text{cond}(\mathbf{A}_r)\,\frac{\|\mathbf{E}\|_F}{\|\mathbf{B}\|_F},$$

where

$$\|\mathbf{X}\|_F = \left(\sum_{i=1}^{m}\sum_{j=1}^{n} x_{ij}^2\right)^{1/2}$$

denotes the **Frobenius norm** of the matrix $\mathbf{X}$. The quantity cond($\mathbf{A}$) is computed by the MATLAB function `cond(A)`. It is the **condition number** of $\mathbf{A}$, formally defined by (1.8), measuring the possible magnification of the relative error in $\mathbf{E}$ in producing the solution $\mathbf{X}_{\text{naïve}}$.

For the test problem in Challenge 1 and different values of the image size, use this relation to determine the maximum allowed value of $\|\mathbf{E}\|_F$ such that the relative error in the naïve reconstruction is guaranteed to be less than 5%.

1.4 Deblurring Using a General Linear Model

Underlying all material in this book is the assumption that the blurring, i.e., the operation of going from the sharp image to the blurred image, is *linear*. As usual in the physical sciences, this assumption is made because in many situations the blur is indeed linear, or at least well approximated by a linear model. An important consequence of the assumption

POINTER. Our basic assumption is that we have a linear blurring process. This means that if $\mathbf{B}_1$ and $\mathbf{B}_2$ are the blurred images of the exact images $\mathbf{X}_1$ and $\mathbf{X}_2$, then $\mathbf{B} = \alpha\,\mathbf{B}_1 + \beta\,\mathbf{B}_2$ is the image of $\mathbf{X} = \alpha\,\mathbf{X}_1 + \beta\,\mathbf{X}_2$. When this is the case, then there exists a large matrix $\mathbf{A}$ such that $\mathbf{b} = \text{vec}(\mathbf{B})$ and $\mathbf{x} = \text{vec}(\mathbf{X})$ are related by the equation

$$\mathbf{A}\,\mathbf{x} = \mathbf{b}.$$

The matrix $\mathbf{A}$ represents the blurring that is taking place in the process of going from the exact to the blurred image. The equation $\mathbf{A}\,\mathbf{x} = \mathbf{b}$ can often be considered as a discretization of an underlying integral equation; the details can be found in [23].

is that we have a large number of tools from linear algebra and matrix computations at our disposal. The use of linear algebra in image reconstruction has a long history, and goes back to classical works such as the book by Andrews and Hunt [1].

In order to handle a variety of applications, we need a blurring model somewhat more general than that in (1.1). The key to obtaining this general linear model is to rearrange the elements of the images $\mathbf{X}$ and $\mathbf{B}$ into column vectors by stacking the columns of these images into two long vectors $\mathbf{x}$ and $\mathbf{b}$, both of length $N = m\,n$. The mathematical notation for this operator is vec, i.e.,

$$\mathbf{x} = \text{vec}(\mathbf{X}) = \begin{bmatrix} \mathbf{x}_1 \\ \vdots \\ \mathbf{x}_n \end{bmatrix} \in \mathbb{R}^N, \qquad \mathbf{b} = \text{vec}(\mathbf{B}) = \begin{bmatrix} \mathbf{b}_1 \\ \vdots \\ \mathbf{b}_n \end{bmatrix} \in \mathbb{R}^N.$$

Since the blurring is assumed to be a linear operation, there must exist a large blurring matrix $\mathbf{A} \in \mathbb{R}^{N \times N}$ such that $\mathbf{x}$ and $\mathbf{b}$ are related by the linear model

$$\mathbf{A}\,\mathbf{x} = \mathbf{b}, \tag{1.4}$$

and this is our fundamental image blurring model. For now, assume that $\mathbf{A}$ is known; we will explain how it can be constructed from the imaging system in Chapter 3, and also discuss the precise structure of the matrix in Chapter 4.

For our linear model, the naïve approach to image deblurring is simply to solve the linear algebraic system in (1.4), but from the previous section, we expect failure. Let us now explain why.

We repeat the computation from the previous section, this time using the general formulation in (1.4). Again let $\mathbf{B}_{\text{exact}}$ and $\mathbf{E}$ be, respectively, the noise-free blurred image and the noise image, and define the corresponding vectors

$$\mathbf{b}_{\text{exact}} = \text{vec}(\mathbf{B}_{\text{exact}}) = \mathbf{A}\,\mathbf{x}, \qquad \mathbf{e} = \text{vec}(\mathbf{E}).$$

Then the noisy recorded image $\mathbf{B}$ is represented by the vector

$$\mathbf{b} = \mathbf{b}_{\text{exact}} + \mathbf{e},$$

and consequently the naïve solution is given by

$$\mathbf{x}_{\text{naïve}} = \mathbf{A}^{-1}\mathbf{b} = \mathbf{A}^{-1}\mathbf{b}_{\text{exact}} + \mathbf{A}^{-1}\mathbf{e} = \mathbf{x} + \mathbf{A}^{-1}\mathbf{e}, \tag{1.5}$$

POINTER. Good presentations of the SVD can be found in the books by Björck [4], Golub and Van Loan [18], and Stewart [57].

where the term $\mathbf{A}^{-1}\mathbf{e}$ is the inverted noise. Equation (1.3) in the previous section is a special case of this equation. The important observation here is that the deblurred image consists of two components: the first component is the exact image, and the second component is the inverted noise. If the deblurred image looks unacceptable, it is because the inverted noise term contaminates the reconstructed image.

Important insight about the inverted noise term can be gained using the singular value decomposition (SVD), which is the tool-of-the-trade in matrix computations for analyzing linear systems of equations. The SVD of a square matrix $\mathbf{A} \in \mathbb{R}^{N\times N}$ is essentially unique and is defined as the decomposition

$$\mathbf{A} = \mathbf{U}\,\mathbf{\Sigma}\,\mathbf{V}^T,$$

where $\mathbf{U}$ and $\mathbf{V}$ are orthogonal matrices, satisfying $\mathbf{U}^T\mathbf{U} = \mathbf{I}_N$ and $\mathbf{V}^T\mathbf{V} = \mathbf{I}_N$, and $\mathbf{\Sigma} = \operatorname{diag}(\sigma_i)$ is a diagonal matrix whose elements σ_i are nonnegative and appear in nonincreasing order,

$$\sigma_1 \geq \sigma_2 \geq \cdots \geq \sigma_N \geq 0.$$

The quantities σ_i are called the singular values, and the rank of $\mathbf{A}$ is equal to the number of positive singular values. The columns $\mathbf{u}_i$ of $\mathbf{U}$ are called the left singular vectors, while the columns $\mathbf{v}_i$ of $\mathbf{V}$ are the right singular vectors. Since $\mathbf{U}^T\mathbf{U} = \mathbf{I}_N$, we see that $\mathbf{u}_i^T\mathbf{u}_j = 0$ if $i \neq j$, and, similarly, $\mathbf{v}_i^T\mathbf{v}_j = 0$ if $i \neq j$.

Assuming for the moment that all singular values are strictly positive, it is straightforward to show that the inverse of $\mathbf{A}$ is given by

$$\mathbf{A}^{-1} = \mathbf{V}\,\mathbf{\Sigma}^{-1}\mathbf{U}^T$$

(we simply verify that $\mathbf{A}^{-1}\mathbf{A} = \mathbf{I}_n$). Since $\mathbf{\Sigma}$ is a diagonal matrix, its inverse $\mathbf{\Sigma}^{-1}$ is also diagonal, with entries $1/\sigma_i$ for $i = 1, \ldots, N$.

Another representation of $\mathbf{A}$ and $\mathbf{A}^{-1}$ is also useful to us:

$$\begin{aligned}
\mathbf{A} &= \mathbf{U}\mathbf{\Sigma}\mathbf{V}^T \\
&= \begin{bmatrix} \mathbf{u}_1 & \cdots & \mathbf{u}_N \end{bmatrix} \begin{bmatrix} \sigma_1 & & \\ & \ddots & \\ & & \sigma_N \end{bmatrix} \begin{bmatrix} \mathbf{v}_1^T \\ \vdots \\ \mathbf{v}_N^T \end{bmatrix} \\
&= \mathbf{u}_1\sigma_1\mathbf{v}_1^T + \cdots + \mathbf{u}_N\sigma_N\mathbf{v}_N^T \\
&= \sum_{i=1}^{N} \sigma_i \mathbf{u}_i\,\mathbf{v}_i^T .
\end{aligned}$$

Similarly,

$$\mathbf{A}^{-1} = \sum_{i=1}^{N} \frac{1}{\sigma_i}\mathbf{v}_i\,\mathbf{u}_i^T .$$

Using this relation, it follows immediately that the naïve reconstruction given in (1.5) can be written as

$$\mathbf{x}_{\text{naïve}} = \mathbf{A}^{-1}\mathbf{b} = \mathbf{V}\,\boldsymbol{\Sigma}^{-1}\mathbf{U}^T\mathbf{b} = \sum_{i=1}^{N} \frac{\mathbf{u}_i^T\mathbf{b}}{\sigma_i}\,\mathbf{v}_i \tag{1.6}$$

and the inverted noise contribution to the solution is given by

$$\mathbf{A}^{-1}\mathbf{e} = \mathbf{V}\,\boldsymbol{\Sigma}^{-1}\mathbf{U}^T\mathbf{e} = \sum_{i=1}^{N} \frac{\mathbf{u}_i^T\mathbf{e}}{\sigma_i}\,\mathbf{v}_i\,. \tag{1.7}$$

In order to understand when this error term dominates the solution, we need to know that the following properties generally hold for image deblurring problems:

- The error components $|\mathbf{u}_i^T\mathbf{e}|$ are small and typically of roughly the same order of magnitude for all i.
- The singular values decay to a value very close to zero. As a consequence the condition number
$$\text{cond}(\mathbf{A}) = \sigma_1/\sigma_N \tag{1.8}$$
is very large, indicating that the solution is very sensitive to perturbation and rounding errors.
- The singular vectors corresponding to the smaller singular values typically represent higher-frequency information. That is, as i increases, the vectors $\mathbf{u}_i$ and $\mathbf{v}_i$ tend to have more sign changes.

The consequence of the last property is that the SVD provides us with basis vectors $\mathbf{v}_i$ for an expansion where each basis vector represents a certain "frequency," approximated by the number of times the entries in the vector change signs.

Figure 1.4 shows images of some of the singular vectors $\mathbf{v}_i$ for the blur of Figure 1.2. Note that each vector $\mathbf{v}_i$ is reshaped into an $m \times n$ array $\mathbf{V}_i$ in such a way that we can write the naïve solution as

$$\mathbf{X}_{\text{naïve}} = \sum_{i=1}^{N} \frac{\mathbf{u}_i^T\mathbf{b}}{\sigma_i}\,\mathbf{V}_i.$$

All the $\mathbf{V}_i$ arrays (except the first) have negative elements and therefore, strictly speaking, they are not images. We see that the spatial frequencies in $\mathbf{V}_i$ increase with the index i.

When we encounter an expansion of the form $\sum_{i=1}^{N} \xi_i\,\mathbf{v}_i$, such as in (1.6) and (1.7), then the ith expansion coefficient ξ_i measures the contribution of $\mathbf{v}_i$ to the result. And since each vector $\mathbf{v}_i$ can be associated with some "frequency," the ith coefficient measures the amount of information of that frequency in our image.

Looking at the expression (1.7), for $\mathbf{A}^{-1}\mathbf{e}$ we see that the quantities $\mathbf{u}_i^T\mathbf{e}/\sigma_i$ are the expansion coefficients for the basis vectors $\mathbf{v}_i$. When these quantities are small in magnitude, the solution has very little contribution from $\mathbf{v}_i$, but when we divide by a small singular value such as σ_N, we greatly magnify the corresponding error component, $\mathbf{u}_N^T\mathbf{e}$, which in turn contributes a large multiple of the high-frequency information contained in $\mathbf{v}_N$ to

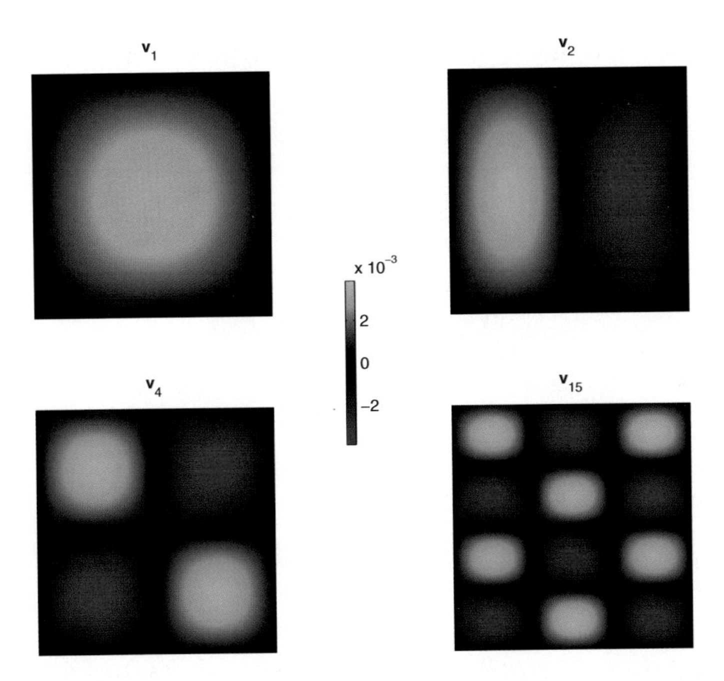

Figure 1.4. *A few of the singular vectors for the blur of the pumpkin image in Figure 1.2. The "images" shown in this figure were obtained by reshaping the $mn \times 1$ singular vectors $\mathbf{v}_i$ into $m \times n$ arrays.*

the computed solution. This is precisely why a naïve reconstruction, such as the one in Figure 1.3, appears as a random image dominated by high frequencies.

Because of this, we might be better off leaving the high-frequency components out altogether, since they are dominated by error. For example, for some choice of $k < N$ we can compute the truncated expansion

$$\mathbf{x}_k = \sum_{i=1}^{k} \frac{\mathbf{u}_i^T \mathbf{b}}{\sigma_i} \mathbf{v}_i \equiv \mathbf{A}_k^\dagger \mathbf{b}$$

in which we have introduced the rank-k matrix

$$\mathbf{A}_k^\dagger = \begin{bmatrix} \mathbf{v}_1 & \cdots & \mathbf{v}_k \end{bmatrix} \begin{bmatrix} \sigma_1 & & \\ & \ddots & \\ & & \sigma_k \end{bmatrix}^{-1} \begin{bmatrix} \mathbf{u}_1^T \\ \vdots \\ \mathbf{u}_k^T \end{bmatrix} = \sum_{i=1}^{k} \frac{1}{\sigma_i} \mathbf{v}_i \mathbf{u}_i^T .$$

Figure 1.5 shows what happens when we replace $\mathbf{A}^{-1}\mathbf{b}$ by $\mathbf{x}_k$ with $k = 800$; this reconstruction is much better than the naïve solution shown in Figure 1.3. We may wonder if a different value for k will produce a better reconstruction!

The truncated SVD expansion for $\mathbf{x}_k$ involves the computation of the SVD of the large $N \times N$ matrix $\mathbf{A}$, and is therefore computationally feasible only if we can find fast algorithms

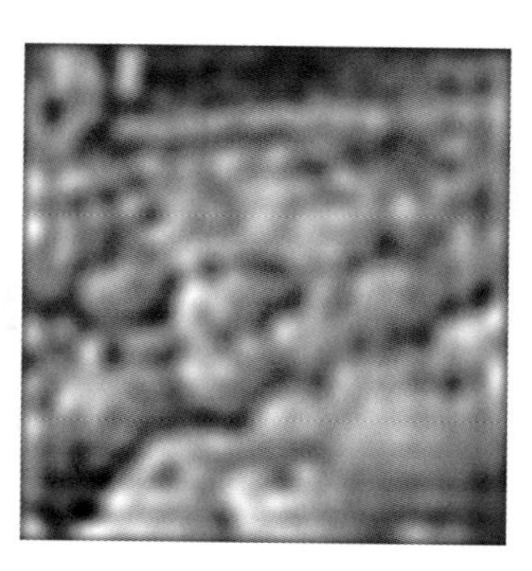

Figure 1.5. *The reconstruction* $\mathbf{x}_k$ *obtained for the blur of the pumpkins of Figure* 1.2 *by using* $k = 800$ *(instead of the full* $k = N = 169744$*).*

to compute the decomposition. Before analyzing such ways to solve our problem, though, it may be helpful to have a brief tutorial on manipulating images in MATLAB, and we present that in the next chapter.

VIP 2. We model the blurring of images as a linear process characterized by a blurring matrix $\mathbf{A}$ and an observed image $\mathbf{B}$, which, in vector form, is $\mathbf{b}$. The reason $\mathbf{A}^{-1}\mathbf{b}$ cannot be used to deblur images is the amplification of high-frequency components of the noise in the data, caused by the inversion of very small singular values of $\mathbf{A}$. Practical methods for image deblurring need to avoid this pitfall.

CHALLENGE 4. For the simple model $\mathbf{B} = \mathbf{A}_c \mathbf{X} \mathbf{A}_r^T + \mathbf{E}$ in Sections 1.2 and 1.3, let us introduce the two rank-k matrices $(\mathbf{A}_c)_k^\dagger$ and $(\mathbf{A}_r)_k^\dagger$, defined similarly to $\mathbf{A}_k^\dagger$. Then for $k < \min(m,n)$ we can define the reconstruction

$$\mathbf{X}_k = (\mathbf{A}_c)_k^\dagger \, \mathbf{B} \left((\mathbf{A}_r)_k^\dagger \right)^T.$$

Use this approach to deblur the image from Challenge 2. Can you find a value of k such that you can read the text?

Chapter 2

Manipulating Images in MATLAB

For the bureaucrat, the world is a mere object to be manipulated by him.
– Karl Marx

We begin this chapter with a recap of how a digital image is stored, and then discuss how to read/load images, how to display them, how to perform arithmetic operations on them, and how to write/save images to files.

2.1 Image Basics

Images can be color, grayscale, or binary (0's and 1's). Color images can use different color models, such as RGB, HSV, and CMY/CMYK. For our purposes, we will use RGB (red, green, blue—the primary colors of light) format for color images, but we will be mainly concerned with grayscale intensity images, which, as shown in Chapter 1, can be thought of simply as two-dimensional arrays (or matrices), where each entry contains the intensity value of the corresponding pixel. Typical grayscales for intensity images can have integer values in the range [0, 255] or [0, 65535], where the lower bound, 0, is black, and the upper bound, 255 or 65535, is white. MATLAB supports each of these formats. In addition,

POINTER. We discuss in this chapter the following MATLAB commands for processing images:

MATLAB		MATLAB IPT
`colormap`	`imformats`	`imshow`
`double`	`importdata`	`rgb2gray`
`imread`	`imwrite`	`mat2gray`
`imfinfo`	`load`	
`image`	`save`	
`imagesc`		

Recall that we use IPT to denote the MATLAB Image Processing Toolbox.

POINTER. An alternative to the RGB format used in this book is CMY (cyan, magenta, and yellow—the subtractive primary colors), often used in the printing industry. Many ink jet printers, for example, use the **CMYK** system, a CMY cartridge and a black one. Another popular color format in image processing is HSV (hue, saturation, value).

MATLAB supports double precision floating point numbers in the interval [0, 1] for pixel values. Since many image processing operations require algebraic manipulation of the pixel values, it may be necessary to allow for noninteger values, so we will convert images to floating point before performing arithmetic operations on them.

2.2 Reading, Displaying, and Writing Images

Here we describe some basics of how to read and display images in MATLAB. The first thing we need is an image. Access to the IPT provides several images we can use; see

```
>> help imdemos/Contents
```

for a full list. In addition, several images can also be downloaded from the book's website. For the examples in this chapter, we use `pumpkins.tif` and `butterflies.tif` from that website.

The command `imfinfo` displays information about the image stored in a data file. For example,

```
>> info = imfinfo('butterflies.tif')
```

shows that the image contained in the file `butterflies.tif` is an RGB image. Doing the same thing for `pumpkins.tif`, we see that this image is a grayscale intensity image.

The command to read images in MATLAB is `imread`. The functions `help` or `doc` describe many ways to use `imread`; here are two simple examples:

```
>> G = imread('pumpkins.tif');
>> F = imread('butterflies.tif');
```

Now use the `whos` command to see what variables we have in our workspace. Notice that both `F` and `G` are arrays whose entries are `uint8` values. This means the intensity values are integers in the range [0, 255]. `F` is a three-dimensional array since it contains RGB information, whereas `G` is a two-dimensional array since it represents only the grayscale intensity values for each pixel.

There are three basic commands for displaying images: `imshow`, `image`, and `imagesc`. In general, `imshow` is preferred, since it renders images more accurately, especially in terms of size and color. However, `imshow` can only be used if the IPT is

POINTER. As part of our software at the book's website, we provide a MATLAB demo script `chapter2demo.m` that performs a step-by-step walk-through of all the commands discussed in this chapter.

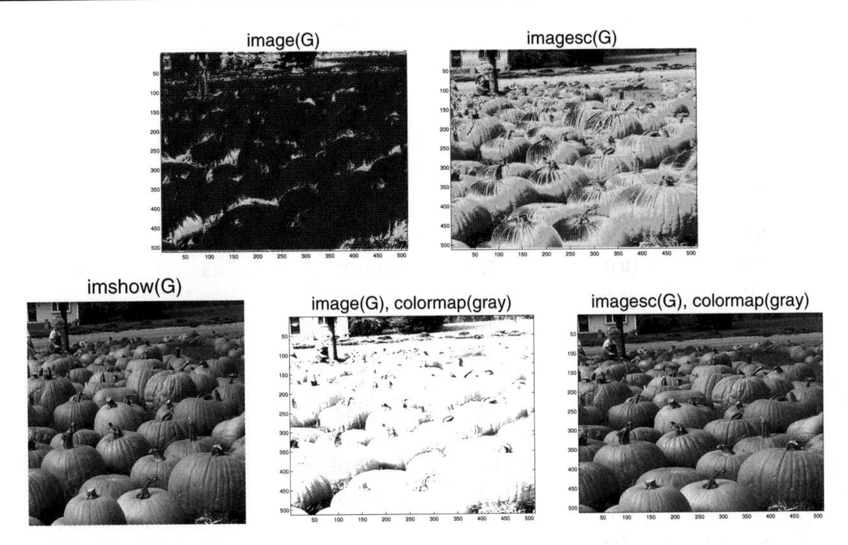

Figure 2.1. *Grayscale pumpkin image displayed by* `imshow`, `image`, *and* `imagesc`. *Only* `imshow` *displays the image with the correct color map and axis ratio.*

available. If this is not the case, then the commands `image` and `imagesc` can be used. We see in Figures 2.1 and 2.2 what happens with each of the following commands:

```
>> figure, imshow(G)
>> figure, image(G)
>> figure, image(G), colormap(gray)
>> figure, imagesc(G)
>> figure, imagesc(G), colormap(gray)
>> figure, imshow(F)
>> figure, image(F)
>> figure, image(F), colormap(gray)
>> figure, imagesc(F)
>> figure, imagesc(F), colormap(gray)
```

In this example, notice that an unexpected rendering may occur. This is especially true for grayscale intensity images, where `image` and `imagesc` display images using a false colormap, unless we explicitly specify gray using the `colormap(gray)` command. In addition, `image` does not always provide a proper scaling of the pixel values. Neither command sets the axis ratio such that the pixels are rendered as squares; this must be done explicitly by the `axis image` command. Thus, if the IPT is not available, we suggest using the `imagesc` command followed by the command `axis image` to get the proper aspect ratio. The tick marks and the numbers on the axes can be removed by the command `axis off`.

To write an image to a file using any of the supported formats we can use the `imwrite` command. There are many ways to use this function, and the online help provides more information. Here we describe only two basic approaches, which will work for converting images from one data format to another, for example, from TIFF to JPEG. This can be done

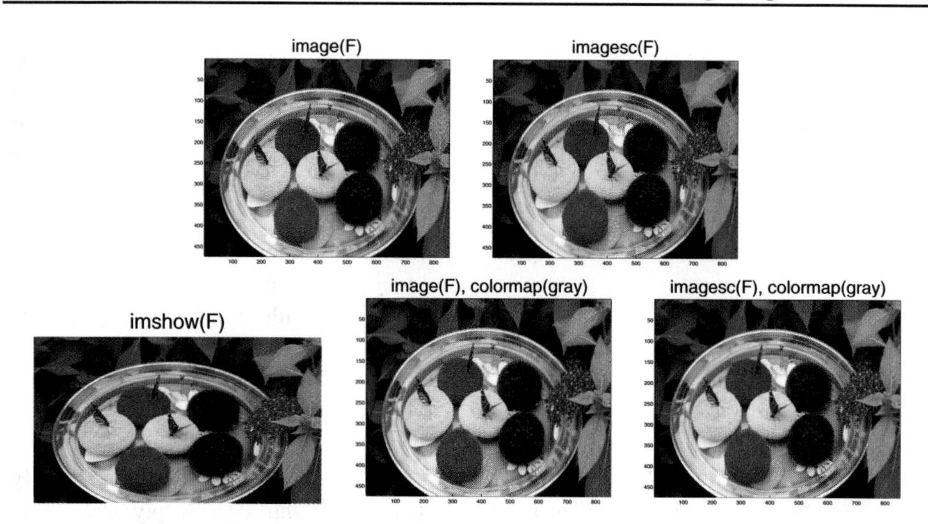

Figure 2.2. *Butterfly image displayed by* `imshow`, `image`, *and* `imagesc`. *Note that* `image` *and* `imagesc` *do not automatically set correct axes, and that the* `gray` *colormap is ignored for color images.*

simply by using `imread` to read an image of one format and `imwrite` to write it to a file of another format. For example,

```
>> G = imread('image.tif');
>> imwrite(G, 'image.jpg');
```

Image data may also be saved in a MAT-file using the `save` command. In this case, if we want to use the saved image in a subsequent MATLAB session, we simply use the `load` command to load the data into the workspace.

2.3 Performing Arithmetic on Images

We've learned the very basics of reading and writing, so now it's time to learn some basics of arithmetic. One important thing we must keep in mind is that most image processing

POINTER. There are many types of image file formats that are used to store images. Currently, the most commonly used formats include

- GIF (Graphics Interchange Format)
- JPEG (Joint Photographic Experts Group)
- PNG (Portable Network Graphics)
- TIFF (Tagged Image File Format)

MATLAB can be used to read and write files with these and many other file formats. The MATLAB command `imformats` provides more information on the supported formats. Note also that MATLAB has its own data format, so images can also be stored using this "MAT-file" format.

software (this includes MATLAB) expects the pixel values (entries in the image arrays) to be in a fixed interval. Recall that typical grayscales for intensity images can have integer values from [0, 255] or [0, 65535], or floating point values in the interval [0, 1]. If, after performing some arithmetic operations, the pixel values fall outside these intervals, unexpected results can occur. Moreover, since our goal is to operate on images with mathematical methods, integer representation of images can be limiting. For example, if we multiply an image by a noninteger scalar, then the result contains entries that are nonintegers. Of course, we can easily convert these to integers by, say, rounding. If we are only doing one arithmetic operation, then this approach may be appropriate; the IPT provides basic image operations such as scaling.

To experiment with arithmetic, we first read in an image:

```
>> G = imread('pumpkins.tif');
```

For the algorithms discussed later in this book, we need to understand how to algebraically manipulate the images; that is, we want to be able to add, subtract, multiply, and divide images. Unfortunately, standard MATLAB commands such as $+$, $-$, $*$, and / do not always work for images. For example, in older versions of MATLAB (e.g., version 6.5), if we attempt the simple command

```
>> G + 10;
```

then we get an error message. The + operator does not work for `uint8` variables! Unfortunately, most images stored as TIFF, JPEG, etc., are either `uint8` or `uint16`, and standard arithmetic operations may not work on these types of variables.

To get around this problem, the IPT has functions such as `imadd`, `imsubtract`, `immultiply`, and `imdivide` that can be used specifically for image operations. However, we will not use these operations.

Our algorithms require *many* arithmetic operations, and working in 8- or 16-bit arithmetic can lead to significant loss of information. Therefore, we adopt the convention of converting the initial image to double precision, operating upon it, and then converting back to the appropriate format when we are ready to display or write an image to a data file.

In working with grayscale intensity images, the main conversion function we need is `double`. It is easy to use

```
>> Gd = double(G);
```

Use the `whos` command to see what variables are contained in the workspace. Notice that `Gd` requires significantly more memory, but now we are not restricted to working only with integers, and standard arithmetic operations like $+$, $-$, $*$, and / all work in a predictable manner.

> **VIP 3.** Before performing arithmetic operations on a grayscale intensity image, use the MATLAB command `double` to convert the pixel values to double precision, floating point numbers.

In some cases, we may want to convert color images to grayscale intensity images. This can be done by using the command `rgb2gray`. Then if we plan to use arithmetic operations on these images, we need to convert to double precision. For example,

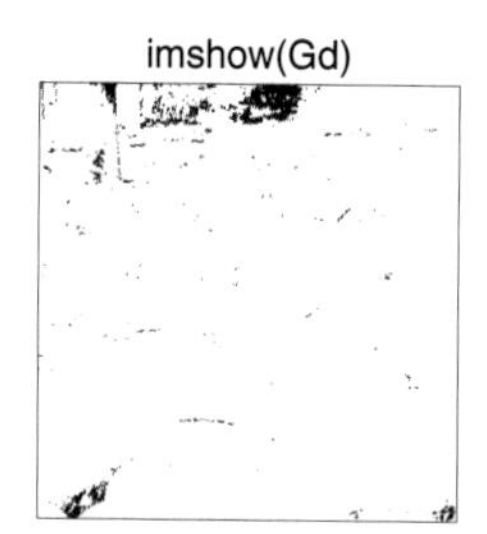

Figure 2.3. *The "double precision" version of the pumpkin image displayed using* `imshow(Gd)` *(left) and* `imagesc(Gd)` *(right).*

```
>> F = imread('butterflies.tif');
>> Fg = rgb2gray(F);
>> Fd = double(Fg);
```

It is not, in general, a good idea to change "true color" images to grayscale, since we lose information.

In any case, once the image is converted to a double precision array, we can use any of MATLAB's array operations on it. For example, to determine the size of the image and the range of its intensities, we can use

```
>> size(Fd)
>> max(Fd(:))
>> min(Fd(:))
```

2.4 Displaying and Writing Revisited

Now that we can perform arithmetic on images, we need to be able to display our results. Note that `Gd` requires more storage space than `G` for the entries in the array, although the values are really the same—look at the values `Gd(200,200)` and `G(200,200)`. But try to display the image `Gd` using the two recommended commands,

```
>> figure, imshow(Gd)
>> figure, imagesc(Gd), axis image, colormap(gray)
```

and we observe that something unusual has occurred when using `imshow`, as shown in Figure 2.3. To understand the problem here, we need to understand how `imshow` works.

- When the input image has `uint8` entries, `imshow` expects the values to be integers in the range 0 (black) to 255 (white).

- When the input image has `uint16` entries, it expects the values to be integers in the range 0 (black) to 65535 (white).

- When the input image has double precision entries, it expects the values to be in the range 0 (black) to 1 (white).

If some entries are not in range, truncation is performed; entries less than 0 are set to 0 (black), and entries larger than the upper bound are set to the white value. Then the image is displayed. The array Gd has entries that range from 0 to 255, but they are double precision. So, before displaying the image, all entries greater than 1 are set to 1, resulting in an image that has only pure black and pure white pixels. We can get around this in two ways. The first is to tell imshow that the max (white) and min (black) are different from 0 and 1 as follows:

```
>> imshow(Gd, [0, 255])
```

Of course this means we need to know the max and min values in the array. If we say

```
>> imshow(Gd, [ ])
```

then imshow finds the max and min values in the array, scales to [0, 1], and then displays.

The other way to fix this scaling problem is to rescale Gd into an array with entries in [0, 1], and then display. This can be done as follows:

```
>> Gds = mat2gray(Gd);
>> imshow(Gds)
```

Probably the most common way we will use imshow is imshow(Gd, [ ]), since it will give consistent results, even if the scaling is already in the interval [0, 1].

> **VIP 4.** If the IPT is available, use the command imshow(G, [ ]) to display image G. If the IPT is not available, use imagesc(G) followed by axis image. To display grayscale intensity images, follow these commands by the command colormap(gray). The tick marks and numbers on the axes can be removed by axis off.

This scaling problem must also be considered when writing images to one of the supported file formats using imwrite. In particular, if the image array is stored as double precision, we should first use the mat2gray command to scale the pixel values to the interval [0, 1]. For example, if X is a double precision array containing grayscale image data, then to save the image as a JPEG or PNG file, we could use

```
>> imwrite(mat2gray(X), 'MyImage.jpg', 'Quality', 100)
>> imwrite(mat2gray(X), 'MyImage.png', 'BitDepth', 16, ...
                        'SignificantBits', 16)
```

If we try this with the double precision pumpkin image, Gd, and then read the image back with imread, we see that the JPEG format saves the image using only 8 bits, while PNG uses 16 bits. If 16 bits is not enough accuracy, and we want to save our images with their full double precision values, then we can simply use the save command. The disadvantages are that the MAT-files are much larger, and they are not easily ported to other applications such as Java programs.

> **VIP 5.** When using the imwrite command to save grayscale intensity images, first use the mat2gray function to properly scale the pixel values.

POINTER. The `importdata` command can be very useful for reading images stored using less popular or more general file formats. For example, images stored using the Flexible Image Transport System (FITS), used by astronomers to archive their images, currently cannot be read using the `imread` command, but can be read using the `importdata` command.

CHALLENGE 5.

Test your understanding of MATLAB's image processing commands. For the grayscale `pumpkins.tif` image, execute the following tasks:

1. Display the image in reverse color (negative); black for white and white for black (see the image on the left above).
2. Display the image in high contrast, replacing pixel values by either 0 or 1 (see the image on the right above).
3. Put the image in a softer focus (i.e., blur it) by replacing each pixel with the average of itself and its eight nearest neighbors.

For the color image `butterflies.tif`, perform these tasks:

4. Display the R, G, and B images separately.
5. Swap the colors: G for the R values, B for G values, and R for B values.
6. Blur the image by applying the averaging technique from task 3 to each of the three colors in the image.
7. Create a grayscale version of the butterflies image by combining 40% of the red channel, 40% of the green channel, and 20% of the blue channel. If you have access to the MATLAB IPT, compare your grayscale image to what is obtained using `rgb2gray`.

Chapter 3

The Blurring Function

Can you keep the deep water still and clear, so it reflects without blurring?
– Lao Tzu

Our main concern in this book is problems in which the significant distortion of the image comes from blurring. We must therefore understand how the blurring matrix $\mathbf{A}$ is constructed, and how we can exploit structure in the matrix when implementing image deblurring algorithms. The latter issues are addressed in the following chapter; in this chapter we describe the components, such as blurring operators, noise, and boundary conditions, that make up the model of the image blurring process. These components provide relations between the original sharp scene and the recorded blurred image and thus provide the information needed to set up a precise mathematical model.

3.1 Taking Bad Pictures

A picture can, of course, be considered "bad" for many reasons. What we have in mind in this book are blurred images, where the blurring comes from some mechanical or physical process that can be described by a linear mathematical model. This model is important, because it allows us to set up an equation whose solution, at least in principle, is the unblurred image.

Everyone who has taken a picture, digital or not, knows what a blurred image looks like—and how to produce a blurred image; we can, for example, defocus the camera's lens!

POINTER. In this chapter we discuss these MATLAB commands:

MATLAB	MATLAB IPT	HNO FUNCTIONS
`conv2`	`fspecial`	`psfDefocus`
`randn`	`imnoise`	`psfGauss`

HNO FUNCTIONS are written by the authors and can be found at the book's website.

POINTER. Technical details about cameras, lenses, and CCDs can be found in many books about computer vision, such as [15].

In defocussing, the blurring comes from the camera itself, more precisely from the optical system in the lens.

No matter how hard we try to focus the camera, there are physical limitations in the construction of the lens that prevent us from producing an ideal sharp image. Some of these limitations are due to the fact that light with many different wavelengths (different colors) goes into the camera, and the exact path followed by the light through the lens depends on the wavelength. Cameras of high quality have lens systems that seek to compensate for this as much as possible.

For many pictures, these limitations are not an issue. But in certain situations—such as microscopy—we need to take such imperfections into consideration, that is, to describe them by a mathematical model.

Sometimes the blurring in an image comes from mechanisms outside the camera, and outside the control of the photographer. A good example is motion blur: the object moved during the time that the shutter was open, with the result that the object appears to be smeared in the recorded image. Obviously we obtain precisely the same effect if the camera moved while the shutter was open.

Yet another type of blurring also taking place outside the camera is due to variations in the air that affect the light coming into the camera. These variations are very often caused by turbulence in the air. You may have noticed how the light above a hot surface (e.g., a highway or desert) tends to flicker. This flicker is due to small variations in the optical path followed by the light, due to the air's turbulence when it is heated.

The same type of atmospheric turbulence also affects images of the earth's surface taken by astronomical telescopes. In spite of many telescopes being located at high altitudes, where air is thin and turbulence is less pronounced, this still causes some blur in the astronomical images. Precisely the same mechanism can blur images taken from a satellite.

The above discussion illustrates just some of the many causes of blurred images.

3.2 The Matrix in the Mathematical Model

Blurred images are visually unappealing, and many photo editing programs for digital image manipulation contain basic tools for enhancing the images, e.g., by "sharpening" the contours in the image. Although these techniques can be useful for mild blurs, they cannot overcome severe blurring that can occur in many important applications. The aim of this book is to describe more sophisticated approaches that can be used for these difficult problems.

POINTER. Blurring in images can arise from many sources, such as limitations of the optical system, camera and object motion, astigmatism, and environmental effects. Examples of blurring, and applications in which they arise, can be found in many places; see, for example, Andrews and Hunt [1], Bertero and Boccacci [3], Jain [31], Lagendijk and Biemond [37], and Roggemann and Welsh [49].

POINTER. Point sources and PSFs are often generated experimentally. What approximates a point source depends on the application. For example, in atmospheric imaging, the point source can be a single bright star [25]. In microscopy, though, the point source is typically a fluorescent microsphere having a diameter that is about half the diffraction limit of the lens [11].

As mentioned in Chapter 1, we take a model-based approach to image deblurring. That is, we assume that the blurring can be described by a mathematical model, and we use this model to reconstruct a sharper, visually more appealing image. Since the key ingredient is the blurring model, we shall take a closer look at the formulation of this model.

We recall that a grayscale image is just an array of dimension $m \times n$ whose elements, the pixels, represent the light intensity. We also recall from Section 1.4 that we can arrange these pixels into a vector of length mn. When we refer to the recorded blurred image, we use the matrix notation $\mathbf{B}$ when referring to the image array, and $\mathbf{b} = \text{vec}(\mathbf{B})$ when referring to the vector representation.

We can also imagine the existence of an exact image, which is the image we would record if the blurring and noise were not present. This image represents the ideal scene that we are trying to capture. Throughout, we make the assumption that this ideal image has the same dimensions as the recorded image, and we refer to it as either the $m \times n$ array $\mathbf{X}$ or the vector $\mathbf{x} = \text{vec}(\mathbf{X})$. Moreover, we will think of the recorded image $\mathbf{b}$ as the blurred version of the ideal image $\mathbf{x}$.

In the linear model, there exists a large matrix $\mathbf{A}$ of dimensions $N \times N$, with $N = mn$, such that $\mathbf{b}$ and $\mathbf{x}$ are related by the equation

$$\mathbf{A}\,\mathbf{x} = \mathbf{b}.$$

The matrix $\mathbf{A}$ represents the blurring that is taking place in the process of going from the exact to the blurred image.

At this stage we know there is such a matrix, but how do we get it? Imagine the following experiment. Suppose we take the exact image to be all black, except for a single bright pixel. If we take a picture of this image, then the blurring operation will cause the single bright pixel to be spread over its neighboring pixels, as illustrated in Figure 3.1. For obvious reasons, the single bright pixel is called a point source, and the function that describes the blurring and the resulting image of the point source is called the point spread function (PSF).

Mathematically, the point source is equivalent to defining an array of all zeros, except a single pixel whose value is 1. That is, we set $\mathbf{x} = \mathbf{e}_i$ to be the ith unit vector,[1] which consists of all zeros except the ith entry, which is 1. The process of taking a picture of this true image is equivalent to computing

$$\mathbf{A}\,\mathbf{e}_i = \mathbf{A}(:,i) = \text{column } i \text{ of } \mathbf{A}.$$

Clearly, if we repeat this process for all unit vectors $\mathbf{e}_i$ for $i = 1, \dots, N$, then in principle we have obtained complete information about the matrix $\mathbf{A}$. In the next section, we explore

[1] The use of $\mathbf{e}_i$ as the ith column of the identity matrix is common in the mathematical literature. It is, however, slightly inconsistent with the notation used in this book since it may be confused with the ith column of the error, $\mathbf{E}$. We have attempted to minimize such inconsistencies.

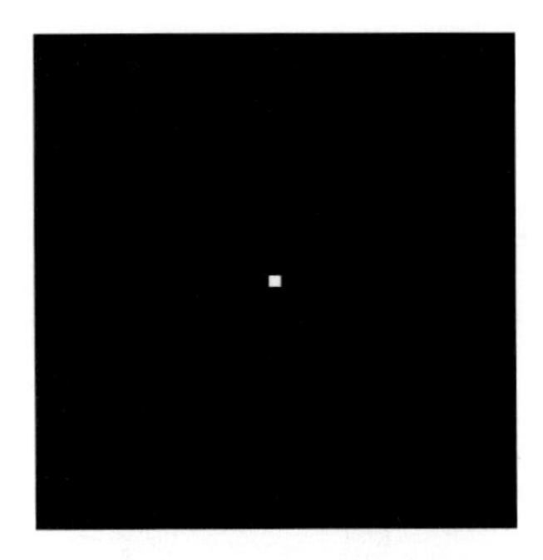

Figure 3.1. *Left: a single bright pixel, called a point source. Right: the blurred point source, called a point spread function.*

alternatives to performing this meticulous task. It might seem that this is all we need to know about the blurring process, but in Sections 3.3 and 3.5 we demonstrate that because we can only see a finite region of a scene that extends forever in all directions, some information is lost in the construction of the matrix **A**. In the next chapter we demonstrate how our deblurring algorithms are affected by the treatment of these boundary conditions.

VIP 6. The blurring matrix **A** is determined from two ingredients: the PSF, which defines how each pixel is blurred, and the boundary conditions, which specify our assumptions on the scene just outside our image.

3.3 Obtaining the PSF

We can learn several important properties of the blurring process by looking at pictures of various PSFs. Consider, for example, Figure 3.2, which shows several PSFs in various locations within the image borders. Here the images are 120×120, and the unit vectors $\mathbf{e}_i$ used to construct these PSFs correspond to the indices $i = 3500, 7150$, and 12555.

In this example—and many others—the light intensity of the PSF is confined to a small area around the center of the PSF (the pixel location of the point source), and outside a certain radius, the intensity is essentially zero. In our example, the PSF is zero 15 pixels from the center. In other words, the blurring is a local phenomenon. Furthermore, if we assume that the imaging process captures all light, then the pixel values in the PSF must sum to 1.

In the example shown here, a careful examination reveals that the PSF is the same regardless of the location of the point source. When this is the case, we say that the blurring is spatially invariant. This is not always the case, but it happens so often that throughout the book we assume spatial invariance.

As a consequence of this linear and local nature of the blurring, to conserve storage we can often represent the PSF using an array **P** of much smaller dimension than the blurred image. (The upper right image in Figure 3.2 has size 31×31.) We refer to **P** as the PSF array.

We remark, though, that many of our deblurring algorithms require that the PSF array

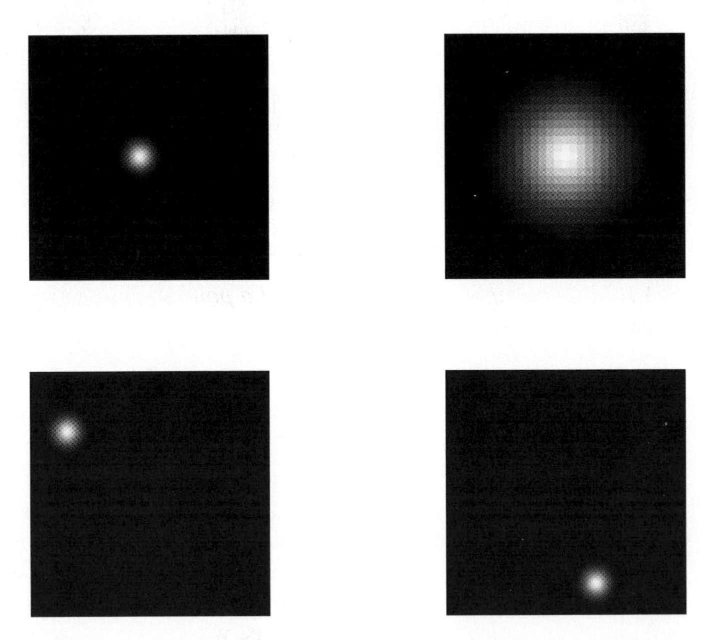

Figure 3.2. *Top: the blurred image of a single pixel (left), and a zoom on the blurred spot (right). Bottom: two blurred images of single pixels near the edges.*

be the same size as the blurred image. In this case the small PSF array is embedded in a larger array of zeros; this process is often referred to as "zero padding" and will be discussed in further detail in Chapter 4.

In some cases the PSF can be described analytically, and thus **P** can be constructed from a function, rather than through experimentation. Consider, for example, horizontal motion blur, which smears a point source into a line. If the line covers r pixels—over which the light is distributed—then the magnitude of each nonzero element in the PSF array is r^{-1}. The same is true for vertical motion blur. An example of the PSF array for horizontal motion is shown in Figure 3.3.

In other cases, knowledge of the physical process that causes the blur provides an explicit formulation of the PSF. When this is the case, the elements of the PSF array are given by a precise mathematical expression. For example, the elements p_{ij} of the PSF array for out-of-focus blur are given by

$$p_{ij} = \begin{cases} 1/(\pi r^2) & \text{if } (i-k)^2 + (j-\ell)^2 \le r^2, \\ 0 & \text{elsewhere,} \end{cases} \tag{3.1}$$

where (k, ℓ) is the center of **P**, and r is the radius of the blur.

The PSF for blurring caused by atmospheric turbulence can be described as a two-dimensional Gaussian function [31, 49], and the elements of the unscaled PSF array are given by

$$p_{ij} = \exp\left(-\frac{1}{2} \begin{bmatrix} i-k \\ j-\ell \end{bmatrix}^T \begin{bmatrix} s_1^2 & \rho^2 \\ \rho^2 & s_2^2 \end{bmatrix}^{-1} \begin{bmatrix} i-k \\ j-\ell \end{bmatrix} \right), \tag{3.2}$$

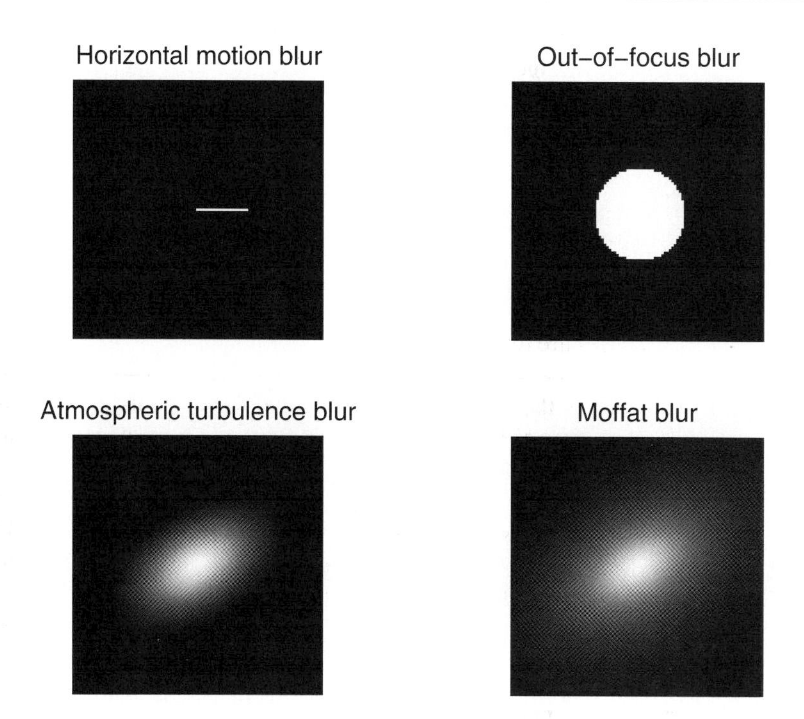

Figure 3.3. *Examples of four PSFs. In all four cases the center of the PSF coincides with the center of the PSF array.*

where the parameters s_1, s_2, and ρ determine the width and the orientation of the PSF, which is centered at element (k, ℓ) in $\mathbf{P}$. Note that one should always scale $\mathbf{P}$ such that its elements sum to 1. The Gaussian function decays exponentially away from the center, and it is reasonable to truncate the values in the PSF array when they have decayed, say, by a factor of 10^4 or 10^8.

The PSF of an astronomical telescope is often modeled by the so-called Moffat function [41], and for this PSF the elements of the unscaled PSF array are given by

$$p_{ij} = \left(1 + \begin{bmatrix} i-k \\ j-\ell \end{bmatrix}^T \begin{bmatrix} s_1^2 & \rho^2 \\ \rho^2 & s_2^2 \end{bmatrix}^{-1} \begin{bmatrix} i-k \\ j-\ell \end{bmatrix}\right)^{-\beta}. \tag{3.3}$$

Similar to the Gaussian PSF for atmospheric turbulence, the parameters s_1, s_2, and ρ determine the width and the orientation of the PSF, and $\mathbf{P}$ should be scaled such that its elements sum to 1. The additional positive parameter β controls the decay of the PSF, which is asymptotically slower than that of the Gaussian PSF.

If $\rho = 0$ in the formulas for the Gaussian blur and Moffat blur, then the PSFs are symmetric along the vertical and horizontal axes, and the formulas take the simpler forms

$$p_{ij} = \exp\left(-\frac{1}{2}\left(\frac{(i-k)}{s_1}\right)^2 - \frac{1}{2}\left(\frac{(j-\ell)}{s_2}\right)^2\right)$$

POINTER. MATLAB's IPT includes a function `fspecial` that computes PSF arrays for motion blur, out-of-focus blur, and Gaussian blur. The functions `psfDefocus` and `psfGauss` can also be found at the book's website.

and

$$p_{ij} = \left(1 + \left(\frac{(i-k)}{s_1}\right)^2 + \left(\frac{(j-\ell)}{s_2}\right)^2\right)^{-\beta}.$$

If also $s_1 = s_2$, then the PSFs are rotationally symmetric.

VIP 7. The PSF array $\mathbf{P}$ is the image of a single white pixel, and its dimensions are usually much smaller than those of $\mathbf{B}$ and $\mathbf{X}$. If the blurring is local and spatially invariant, then $\mathbf{P}$ contains all information about the blurring throughout the image.

Once the PSF array is specified, we can always construct the big blurring matrix $\mathbf{A}$ one column at a time by simply placing the elements of $\mathbf{P}$ in the appropriate positions, leaving zeros elsewhere in the column. In the next chapter, we shall see how the locality and the spatial invariance impose a special structure on the matrix $\mathbf{A}$, which saves us this cumbersome work.

If we want to compute the blurred image $\mathbf{B}$ one pixel at a time (given the sharp image $\mathbf{X}$), then we need to compute

$$b_i = \mathbf{e}_i^T \mathbf{b} = \mathbf{e}_i^T \mathbf{A}\mathbf{x} = \mathbf{A}(i,:)\mathbf{x}.$$

Hence we need to work with the rows of $\mathbf{A}$—not the columns—to compute each pixel in the blurred image as a weighted sum (or average) of the corresponding element and its neighbors in the sharp image. The weights are the elements of the rows in $\mathbf{A}$. Alternatively, we can use the fact that the weights are also given by the pixel values of the PSF array $\mathbf{P}$, and the weighted sum operation is known in mathematics and image processing as a two-dimensional convolution.

CHALLENGE 6. Write a MATLAB function `psfMoffat` (similar to our functions `psfDefocus` and `psfGauss`) with the call

```
P = psfMoffat(dim, s, beta)
```

that computes the PSF array `P` for Moffat blur, using (3.3) for the case with $s_1 = s_2 =$ `s`, $\rho = 0$, and $\beta =$ `beta`. The PSF array should have dimensions `dim(1)` × `dim(2)`, and the center of the PSF should be located at the center of the array.

POINTER. MATLAB has a built-in function, `conv2`, that can be used to form the convolution of a PSF image and a true image—in other words, to artificially blur an image. The function is computationally efficient if the PSF image array has small dimensions, but for larger arrays it is better to use an approach based on the fast Fourier transform; see Chapter 4 for details.

CHALLENGE 7. In this book the convolution with $\mathbf{P}$ is used as a mathematical model for the blurring in the picture. Explicit convolution with a (typically small) PSF array $\mathbf{P}$ can also be used as a computational device to filter an image, and this challenge illustrates three such applications.

- Noise removal is achieved by averaging each pixel and its nearest neighbors (cf. Challenge 5), typically using one of the following 3×3 PSF arrays:

$$\frac{1}{9}\begin{bmatrix} 1 & 1 & 1 \\ 1 & 1 & 1 \\ 1 & 1 & 1 \end{bmatrix}, \qquad \frac{1}{10}\begin{bmatrix} 1 & 1 & 1 \\ 1 & 2 & 1 \\ 1 & 1 & 1 \end{bmatrix}, \qquad \frac{1}{16}\begin{bmatrix} 1 & 2 & 1 \\ 2 & 4 & 2 \\ 1 & 2 & 1 \end{bmatrix}.$$

  Each of these low-pass filters is normalized to reproduce a constant image. Load the images `pumpkinsnoisy1.tif` and `pumpkinsnoisy2.tif` and try to remove the noise by means of these filters.

- Edge detection can be achieved by means of a high-pass filter that damps the low frequencies in the image while maintaining the high frequencies. Examples of such filters are the following 3×3 PSF arrays:

$$\begin{bmatrix} 0 & -1 & 0 \\ -1 & 4 & -1 \\ 0 & -1 & 0 \end{bmatrix}, \qquad \begin{bmatrix} -1 & -1 & -1 \\ -1 & 8 & -1 \\ -1 & -1 & -1 \end{bmatrix}, \qquad \begin{bmatrix} 1 & -2 & 1 \\ -2 & 4 & -2 \\ 1 & -2 & 1 \end{bmatrix}.$$

  All three filters produce zero when applied to a constant image. Load the image `pumpkins.tif` and then try the three filters.

- Edge enhancement is achieved by adding some amount of the high-pass filtered image to the original image, resulting in an image that appears "sharper." (This is not deblurring.) Test this approach on the two images `pumpkinsblurred1.tif` and `pumpkinsblurred2.tif`, using each of the three high-pass filters. Remember to use the same color axis on the original and the enhanced image.

There are many other linear and nonlinear filters used in image processing, such as the median and Sobel filters; see, for example, [31] and the MATLAB IPT functions `medfilt2` and `fspecial`.

3.4 Noise

In addition to blurring, observed images are usually contaminated with noise.[2] Noise can arise from several sources and can be linear, nonlinear, multiplicative, and additive. In this book we consider a common additive noise model that is used for CCD arrays; see, for example, [2, 55, 56]. In this model, noise comes essentially from the following three sources:

- Background photons, from both natural and artificial sources, cause noise to corrupt each pixel value measured by the CCD array. This kind of noise is typically modeled

[2] Images can be corrupted by other defects. For example, "bad pixels" occur if the CCD array used to collect the image has broken elements. We do not consider such errors in this book.

by a Poisson process, with a fixed Poisson parameter, and is thus often referred to as Poisson noise. This kind of noise can take only positive values. In MATLAB, Poisson noise can be generated with the function `imnoise` from the IPT, or with the function `poissrnd` from the Statistics Toolbox.

- The CCD electronics and the analog-to-digital conversion of measured voltages result in readout noise. Readout noise is usually assumed to consist of independent and identically distributed random values; this is called white noise.[3] The noise is further assumed to be drawn from a Gaussian (i.e., normal) distribution with mean 0 and a fixed standard deviation proportional to the amplitude of the noise. Such random errors are often called Gaussian white noise. In MATLAB, Gaussian white noise can be generated using the built-in function `randn`.

- The analog-to-digital conversion also results in quantization error, when the signal is represented by a finite (small) number of bits. Quantization error can be approximated by uniformly distributed white noise whose standard deviation is inversely proportional to the number of bits used.

In our linear algebra notation, we can describe the inclusion of additive noise to the blurred image as follows:

$$\mathbf{B} = \mathbf{B}_{\text{exact}} + \mathbf{E},$$

where $\mathbf{E}$ is an $m \times n$ array containing, for example, elements from a Poisson or Gaussian distribution (or a sum of both). For example, Gaussian white noise with standard deviation 0.01 is generated with the MATLAB command `E = 0.01*randn(m,n)`. Similarly, uniformly distributed white noise with standard deviation 1 is generated with the command `sqrt(3)*(2*rand(m,n)-1)`, so we can use `0.01*sqrt(3)*(2*rand(m,n)-1)` to obtain uniformly distributed white noise with standard deviation 0.01.

In some cases it is possible to assume that the Poisson noise can be approximated well by Gaussian white noise [2], and that the quantization noise is negligible. We can thus generate fairly accurate noise models by simply using MATLAB's built-in `randn` function.

3.5 Boundary Conditions

The discussion of the PSF in Section 3.3 reveals a potential difficulty at the boundaries of the image, where information from the exact image "spills over the edge" of the recorded blurred image. Clearly we lose some information that cannot be recovered; see Figure 3.4. Consider a pixel in a blurred image, say b_{ij}, that is near the edge of the picture. Recall that b_{ij} is obtained from a weighted sum of pixel x_{ij} and its neighbors, some of which may be outside the field of view. When we described how the columns of $\mathbf{A}$ could be constructed in Section 3.2, we ignored this behavior at the boundary.

However, a good model for image deblurring must take account of these boundary effects—otherwise the reconstruction will likely contain some unwanted artifacts near the boundary. Such artifacts can easily be seen in the reconstructed pumpkin image in Figure 1.5.

The most common technique for dealing with this missing information at the boundary is to make certain assumptions about the behavior of the sharp image outside the boundary.

[3]The term "white" is used because these random errors have certain spectral similarities with white light.

Figure 3.4. *The PSF "spills over the edge" at the image boundary (the yellow line), so that scene values outside the image affect what is recorded.*

When these assumptions are used in the blurring model, we say that we impose boundary conditions on the reconstruction. We shall describe some boundary conditions that can be expressed in our language of matrix computations. There are many other techniques, such as those based on extrapolation (e.g., via a statistical analysis of the image), beyond the scope of our model.

Our boundary conditions come in different forms. The simplest boundary condition is to assume that the exact image is black (i.e., consists of zeros) outside the boundary. This zero boundary condition can be pictured as embedding the image $\mathbf{X}$ in a larger image:

$$\mathbf{X}_{\text{ext}} = \begin{bmatrix} \mathbf{0} & \mathbf{0} & \mathbf{0} \\ \mathbf{0} & \mathbf{X} & \mathbf{0} \\ \mathbf{0} & \mathbf{0} & \mathbf{0} \end{bmatrix}, \tag{3.4}$$

where the $\mathbf{0}$ submatrices represent a border of zero elements. If we ignore boundary conditions in creating the blurring matrix $\mathbf{A}$ (as in Section 3.2), we implicitly assume zero boundary conditions.

The zero boundary condition is a good choice when the exact image is mostly zero outside the boundary—as is the case for many astronomical images with a black background. Unfortunately, the zero boundary condition has a bad effect on reconstructions of images that are nonzero outside the border. Sometimes we merely get an artificial black border; at other times we compute a reconstructed image with severe "ringing" near the boundary, caused by a large difference in pixel values inside and outside of the border.

Hence we must often use other boundary conditions that impose a more realistic model of the behavior of the image at the boundary but only make use of the information available, i.e., the image within the boundaries.

The periodic boundary condition is frequently used in image processing. This implies that the image repeats itself (endlessly) in all directions. Again we can picture this boundary condition as embedding the image $\mathbf{X}$ in a larger image that consists of replicas of $\mathbf{X}$:

$$\mathbf{X}_{\text{ext}} = \begin{bmatrix} \mathbf{X} & \mathbf{X} & \mathbf{X} \\ \mathbf{X} & \mathbf{X} & \mathbf{X} \\ \mathbf{X} & \mathbf{X} & \mathbf{X} \end{bmatrix}. \tag{3.5}$$

In some applications it is reasonable to use a reflexive boundary condition, which implies that the scene outside the image boundaries is a mirror image of the scene inside the image boundaries. Introducing (with MATLAB notation) the three additional images,

$$\mathbf{X}_{\text{lr}} = \texttt{fliplr}(\mathbf{X}), \qquad \mathbf{X}_{\text{ud}} = \texttt{flipud}(\mathbf{X}), \qquad \mathbf{X}_{\times} = \texttt{fliplr}(\mathbf{X}_{\text{ud}}),$$

we can picture the reflexive boundary condition as embedding the image $\mathbf{X}$ in the following larger image:

$$\mathbf{X}_{\text{ext}} = \begin{bmatrix} \mathbf{X}_{\times} & \mathbf{X}_{\text{ud}} & \mathbf{X}_{\times} \\ \mathbf{X}_{\text{lr}} & \mathbf{X} & \mathbf{X}_{\text{lr}} \\ \mathbf{X}_{\times} & \mathbf{X}_{\text{ud}} & \mathbf{X}_{\times} \end{bmatrix}. \tag{3.6}$$

We illustrate this approach with a 3×3 example,

$$\mathbf{X} = \begin{bmatrix} 1 & 2 & 3 \\ 4 & 5 & 6 \\ 7 & 8 & 9 \end{bmatrix}, \qquad \mathbf{X}_{\text{ext}} = \left[\begin{array}{ccc|ccc|ccc} 9 & 8 & 7 & 7 & 8 & 9 & 9 & 8 & 7 \\ 6 & 5 & 4 & 4 & 5 & 6 & 6 & 5 & 4 \\ 3 & 2 & 1 & 1 & 2 & 3 & 3 & 2 & 1 \\ \hline 3 & 2 & 1 & 1 & 2 & 3 & 3 & 2 & 1 \\ 6 & 5 & 4 & 4 & 5 & 6 & 6 & 5 & 4 \\ 9 & 8 & 7 & 7 & 8 & 9 & 9 & 8 & 7 \\ \hline 9 & 8 & 7 & 7 & 8 & 9 & 9 & 8 & 7 \\ 6 & 5 & 4 & 4 & 5 & 6 & 6 & 5 & 4 \\ 3 & 2 & 1 & 1 & 2 & 3 & 3 & 2 & 1 \end{array}\right].$$

Which boundary condition to use is often dictated by the particular application.

VIP 8. Boundary conditions specify our assumptions on the behavior of the scene outside the boundaries of the given image. In order to obtain a high-quality deblurred image we must choose the boundary conditions appropriately. Ignoring boundary conditions is equivalent to assuming zero boundary conditions.

CHALLENGE 8.

Let $\mathbf{X}$ be the `iogray.tif` image shown above, of size 512×512. Construct the three versions of the extended image $\mathbf{X}_{\text{ext}}$ corresponding to the three types of boundary conditions, and blur each image with the Gaussian (atmospheric turbulence) blur with $s_1 = s_2 = 15$ and $\rho = 0$, using a small PSF array `P` of size 32×32. To compute the PSF array, you can use the call

```
P = psfGauss(32, 15).
```

To perform the blurring, you can use the call

$$\mathbf{B}_{\text{ext}} = \texttt{conv2}(\mathbf{X}_{\text{ext}}, \texttt{P}, \texttt{'same'}).$$

Finally extract as $\mathbf{B}$ the center part of $\mathbf{B}_{\text{ext}}$ such that $\mathbf{B}$ corresponds to $\mathbf{X}$. Which boundary condition provides the best deblurring model, i.e., gives the smallest amount of artifacts at borders of the image $\mathbf{B}$?

Chapter 4

Structured Matrix Computations

The structure will automatically provide the pattern for the action which follows.
– Donald Curtis

Now that we understand the basic components (i.e., PSF, boundary conditions, and noise) of the image blurring model, we are in a position to provide an explicit description of the blurring matrix, $\mathbf{A}$, defined by the linear model,

$$\mathbf{b} = \mathbf{A}\mathbf{x} + \mathbf{e}.$$

The deblurring algorithms in this book use certain orthogonal or unitary decompositions of the matrix $\mathbf{A}$. We have already encountered the SVD,

$$\mathbf{A} = \mathbf{U}\mathbf{\Sigma}\mathbf{V}^T, \tag{4.1}$$

where $\mathbf{U}$ and $\mathbf{V}$ are orthogonal matrices, and $\mathbf{\Sigma}$ is a diagonal matrix. Another useful decomposition is the (unitary) spectral decomposition,

$$\mathbf{A} = \widetilde{\mathbf{U}}\mathbf{\Lambda}\widetilde{\mathbf{U}}^*, \tag{4.2}$$

where $\widetilde{\mathbf{U}}$ is a unitary matrix[4] and $\mathbf{\Lambda}$ is a diagonal matrix containing the eigenvalues of $\mathbf{A}$. While it is possible to compute an SVD for any matrix, the spectral decomposition can be computed if and only if $\mathbf{A}$ is a normal matrix;[5] that is, $\mathbf{A}^*\,\mathbf{A} = \mathbf{A}\,\mathbf{A}^*$. Note also that if $\mathbf{A}$ has real entries, then the elements in the matrices of the SVD will be real, but the elements in the spectral decomposition may be complex. Also, when $\mathbf{A}$ is real, then its eigenvalues (the diagonal elements of $\mathbf{\Lambda}$) are either real or appear in complex conjugate pairs.

[4] A matrix is unitary if $\widetilde{\mathbf{U}}^*\,\widetilde{\mathbf{U}} = \widetilde{\mathbf{U}}\,\widetilde{\mathbf{U}}^* = \mathbf{I}$, where $\widetilde{\mathbf{U}}^* = \mathrm{conj}(\widetilde{\mathbf{U}})^T$ is the complex conjugate transpose of $\widetilde{\mathbf{U}}$.

[5] If the matrix is not normal, but there are N linearly independent eigenvectors, then it is possible to compute the eigendecomposition $\mathbf{A} = \widetilde{\mathbf{U}}\mathbf{\Lambda}\widetilde{\mathbf{U}}^{-1}$, where $\mathbf{U}$ is an invertible (but not unitary) matrix containing N linearly independent eigenvectors of $\mathbf{A}$. In this book we consider only the orthogonal and unitary decompositions given by (4.1) and (4.2).

POINTER. The following functions are used to implement some of the structured matrix computations discussed in this chapter. HNO FUNCTIONS refer to M-files written by the authors, and may be found in Appendix 3 and at the book's website.

MATLAB	MATLAB IPT	HNO FUNCTIONS
`fft2`	`dct2`	`dcts2`
`ifft2`	`idct2`	`idcts2`
`circshift`		`dctshift`
`svds`		`kronDecomp`
`svd`		`padPSF`
`randn`		

In Chapter 1 we used the SVD to briefly investigate the sensitivity of the image deblurring problem. A similar analysis can be done using the spectral decomposition, if $\mathbf{A}$ is normal. Although it may be prohibitively expensive to compute these factorizations for large, generic matrices, efficient approaches exist for certain structured matrices. The purpose of this chapter is to describe various kinds of structured matrices that arise in certain image deblurring problems, and to show how to efficiently compute the SVD or spectral decomposition of such matrices. By efficient we mean that the decomposition can be computed quickly and that storage requirements remain manageable. Structured matrices can often be uniquely represented by a small number of entries, but this does not necessarily mean that there is similar exploitable structure in the matrices $\mathbf{U}$, $\mathbf{V}$, and $\widetilde{\mathbf{U}}$ that make up the SVD and spectral decomposition.

4.1 Basic Structures

For spatially invariant image deblurring problems, the specific structure of the matrix $\mathbf{A}$ depends on the imposed boundary conditions, and can involve *Toeplitz*, *circulant*, and *Hankel* matrices. To simplify the notation, we first describe, in Section 4.1.1, how these structures arise for one-dimensional problems. The extension to two-dimensional problems is straightforward and is given in Section 4.1.2.

4.1.1 One-Dimensional Problems

Recall that by convolving a PSF with a true image, we obtain a blurred image. Convolution is a mathematical operation that can be described as follows. If $p(s)$ and $x(s)$ are continuous functions, then the convolution of p and x is a function b having the form

$$b(s) = \int_{-\infty}^{\infty} p(s-t)x(t)dt.$$

That is, each value of the function $b(s)$ is essentially a weighted average of the values of $x(t)$, where the weights are given by the function p. In order to perform the integration, we must first "flip" the function $p(t)$ to obtain $p(-t)$, and then "shift" to obtain $p(s-t)$.

The discrete version of convolution is a summation over a finite number of terms. That is, pixels of the blurred image are obtained from a weighted sum of the corresponding pixel and its neighbors in the true image. The weights are given by the elements in the PSF array. It is perhaps easier to illustrate the discrete operation for one-dimensional convolution using a small example. Suppose a true image scene and PSF array are given, respectively, by the one-dimensional arrays

$$\begin{bmatrix} w_1 \\ w_2 \\ \hline x_1 \\ x_2 \\ x_3 \\ x_4 \\ x_5 \\ \hline y_1 \\ y_2 \end{bmatrix} \quad \text{and} \quad \begin{bmatrix} p_1 \\ p_2 \\ p_3 \\ p_4 \\ p_5 \end{bmatrix},$$

where w_i and y_i represent pixels in the original scene that are actually outside the field of view. The basic idea of computing the convolution, $\mathbf{b}$, of $\mathbf{x}$ and $\mathbf{p}$, can be summarized as follows:

- Rotate (i.e., flip) the PSF array, $\mathbf{p}$, by 180 degrees, writing its elements from bottom to top.
- Match coefficients of the rotated PSF array with those in $\mathbf{x}$ by placing (i.e., shifting) the center of the PSF array (i.e., the entry corresponding to a shift of zero) over the ith entry in $\mathbf{x}$.
- Multiply corresponding components, and sum to get the ith entry in $\mathbf{b}$.

For example, assuming that p_3 is the center of the PSF array, we have

$$\begin{aligned} b_1 &= p_5 w_1 + p_4 w_2 + p_3 x_1 + p_2 x_2 + p_1 x_3, \\ b_2 &= p_5 w_2 + p_4 x_1 + p_3 x_2 + p_2 x_3 + p_1 x_4, \\ b_3 &= p_5 x_1 + p_4 x_2 + p_3 x_3 + p_2 x_4 + p_1 x_5, \\ b_4 &= p_5 x_2 + p_4 x_3 + p_3 x_4 + p_2 x_5 + p_1 y_1, \\ b_5 &= p_5 x_3 + p_4 x_4 + p_3 x_5 + p_2 y_1 + p_1 y_2. \end{aligned}$$

Thus, convolution can be written as a matrix-vector multiplication,

$$\begin{bmatrix} b_1 \\ b_2 \\ b_3 \\ b_4 \\ b_5 \end{bmatrix} = \begin{bmatrix} p_5 & p_4 & p_3 & p_2 & p_1 & & & & \\ & p_5 & p_4 & p_3 & p_2 & p_1 & & & \\ & & p_5 & p_4 & p_3 & p_2 & p_1 & & \\ & & & p_5 & p_4 & p_3 & p_2 & p_1 & \\ & & & & p_5 & p_4 & p_3 & p_2 & p_1 \end{bmatrix} \begin{bmatrix} w_1 \\ w_2 \\ \hline x_1 \\ x_2 \\ x_3 \\ x_4 \\ x_5 \\ \hline y_1 \\ y_2 \end{bmatrix}. \tag{4.3}$$

It is important to keep in mind that the values w_i and y_i contribute to the observed pixels in the blurred image even though they are outside the field of view. Since we do not know these values, boundary conditions are used to relate them to the pixels x_i within the field of view. Using the various boundary conditions discussed in Section 3.5 we have the following.

- Zero Boundary Conditions. In this case, $w_i = y_i = 0$, and thus (4.3) can be rewritten as

$$\begin{bmatrix} b_1 \\ b_2 \\ b_3 \\ b_4 \\ b_5 \end{bmatrix} = \begin{bmatrix} p_3 & p_2 & p_1 & & \\ p_4 & p_3 & p_2 & p_1 & \\ p_5 & p_4 & p_3 & p_2 & p_1 \\ & p_5 & p_4 & p_3 & p_2 \\ & & p_5 & p_4 & p_3 \end{bmatrix} \begin{bmatrix} x_1 \\ x_2 \\ x_3 \\ x_4 \\ x_5 \end{bmatrix}. \tag{4.4}$$

A matrix whose entries are constant on each diagonal, such as in (4.4), is called a Toeplitz matrix with parameters $\mathbf{p}$.

- Periodic Boundary Conditions. In this case we assume that the true scene is comprised of periodic copies of $\mathbf{x}$, so $w_1 = x_4$, $w_2 = x_5$, $y_1 = x_1$, and $y_2 = x_2$. Thus (4.3) can be rewritten as

$$\begin{bmatrix} b_1 \\ b_2 \\ b_3 \\ b_4 \\ b_5 \end{bmatrix} = \left(\begin{bmatrix} p_3 & p_2 & p_1 & & \\ p_4 & p_3 & p_2 & p_1 & \\ p_5 & p_4 & p_3 & p_2 & p_1 \\ & p_5 & p_4 & p_3 & p_2 \\ & & p_5 & p_4 & p_3 \end{bmatrix} + \begin{bmatrix} & & & p_5 & p_4 \\ & & & & p_5 \\ & & & & \\ p_1 & & & & \\ p_2 & p_1 & & & \end{bmatrix} \right) \begin{bmatrix} x_1 \\ x_2 \\ x_3 \\ x_4 \\ x_5 \end{bmatrix}$$

$$= \begin{bmatrix} p_3 & p_2 & p_1 & p_5 & p_4 \\ p_4 & p_3 & p_2 & p_1 & p_5 \\ p_5 & p_4 & p_3 & p_2 & p_1 \\ p_1 & p_5 & p_4 & p_3 & p_2 \\ p_2 & p_1 & p_5 & p_4 & p_3 \end{bmatrix} \begin{bmatrix} x_1 \\ x_2 \\ x_3 \\ x_4 \\ x_5 \end{bmatrix}. \tag{4.5}$$

A Toeplitz matrix in which each row (and column) is a periodic shift of its previous row (column), such as in (4.5), is called a circulant matrix.

- Reflexive Boundary Conditions. In this case we assume that the true scene immediately outside the field of view is a mirror reflection of the scene within the field of view. Thus $w_1 = x_2$, $w_2 = x_1$, $y_1 = x_5$, and $y_2 = x_4$, and (4.3) can be rewritten as

$$\begin{bmatrix} b_1 \\ b_2 \\ b_3 \\ b_4 \\ b_5 \end{bmatrix} = \left(\begin{bmatrix} p_3 & p_2 & p_1 & & \\ p_4 & p_3 & p_2 & p_1 & \\ p_5 & p_4 & p_3 & p_2 & p_1 \\ & p_5 & p_4 & p_3 & p_2 \\ & & p_5 & p_4 & p_3 \end{bmatrix} + \begin{bmatrix} p_4 & p_5 & & & \\ p_5 & & & & \\ & & & & \\ & & & & p_1 \\ & & & p_1 & p_2 \end{bmatrix} \right) \begin{bmatrix} x_1 \\ x_2 \\ x_3 \\ x_4 \\ x_5 \end{bmatrix}$$

$$= \begin{bmatrix} p_3 + p_4 & p_2 + p_5 & p_1 & & \\ p_4 + p_5 & p_3 & p_2 & p_1 & \\ p_5 & p_4 & p_3 & p_2 & p_1 \\ & p_5 & p_4 & p_3 & p_2 + p_1 \\ & & p_5 & p_4 + p_1 & p_3 + p_2 \end{bmatrix} \begin{bmatrix} x_1 \\ x_2 \\ x_3 \\ x_4 \\ x_5 \end{bmatrix}. \tag{4.6}$$

A matrix whose entries are constant on each antidiagonal is called a Hankel matrix, so the matrix given in (4.6) is a Toeplitz-plus-Hankel matrix.

4.1.2 Two-Dimensional Problems

The convolution operation for two-dimensional images is very similar to the one-dimensional case. In particular, to compute the (i, j) pixel of the convolved image, $\mathbf{B}$, the PSF array is rotated by 180 degrees and matched with pixels in the true scene, $\mathbf{X}$, by placing the center of the rotated PSF array over the (i, j) pixel in $\mathbf{X}$. Corresponding components are multiplied, and the results summed to compute b_{ij}. For example, let

$$\mathbf{X} = \begin{bmatrix} x_{11} & x_{12} & x_{13} \\ x_{21} & x_{22} & x_{23} \\ x_{31} & x_{32} & x_{33} \end{bmatrix}, \quad \mathbf{P} = \begin{bmatrix} p_{11} & p_{12} & p_{13} \\ p_{21} & p_{22} & p_{23} \\ p_{31} & p_{32} & p_{33} \end{bmatrix}, \quad \mathbf{B} = \begin{bmatrix} b_{11} & b_{12} & b_{13} \\ b_{21} & b_{22} & b_{23} \\ b_{31} & b_{32} & b_{33} \end{bmatrix},$$

with p_{22} the center of the PSF array. Then the element b_{22} in the center of $\mathbf{B}$ is given by

$$\begin{aligned} b_{22} = {} & p_{33} \cdot x_{11} + p_{32} \cdot x_{12} + p_{31} \cdot x_{13} \\ & + p_{23} \cdot x_{21} + p_{22} \cdot x_{22} + p_{21} \cdot x_{23} \\ & + p_{13} \cdot x_{31} + p_{12} \cdot x_{32} + p_{11} \cdot x_{33}. \end{aligned}$$

For all other elements in this 3×3 example, we must make use of the boundary conditions. In particular, if we assume zero boundary conditions, then the element b_{21} at the border of $\mathbf{B}$ is given by

$$\begin{aligned} b_{21} = {} & p_{33} \cdot 0 + p_{32} \cdot x_{11} + p_{31} \cdot x_{12} \\ & + p_{23} \cdot 0 + p_{22} \cdot x_{21} + p_{21} \cdot x_{22} \\ & + p_{13} \cdot 0 + p_{12} \cdot x_{31} + p_{11} \cdot x_{32}. \end{aligned}$$

By carrying out this exercise for all the elements of $\mathbf{B}$, it is straightforward to show that for zero boundary conditions, $\mathbf{b} = \text{vec}(\mathbf{B})$ and $\mathbf{x} = \text{vec}(\mathbf{X})$ are related by

$$\begin{bmatrix} b_{11} \\ b_{21} \\ b_{31} \\ b_{12} \\ b_{22} \\ b_{32} \\ b_{13} \\ b_{23} \\ b_{33} \end{bmatrix} = \left[\begin{array}{ccc|ccc|ccc} p_{22} & p_{12} & & p_{21} & p_{11} & & & & \\ p_{32} & p_{22} & p_{12} & p_{31} & p_{21} & p_{11} & & & \\ & p_{32} & p_{22} & & p_{31} & p_{21} & & & \\ \hline p_{23} & p_{13} & & p_{22} & p_{12} & & p_{21} & p_{11} & \\ p_{33} & p_{23} & p_{13} & p_{32} & p_{22} & p_{12} & p_{31} & p_{21} & p_{11} \\ & p_{33} & p_{23} & & p_{32} & p_{22} & & p_{31} & p_{21} \\ \hline & & & p_{23} & p_{13} & & p_{22} & p_{12} & \\ & & & p_{33} & p_{23} & p_{13} & p_{32} & p_{22} & p_{12} \\ & & & & p_{33} & p_{23} & & p_{32} & p_{22} \end{array}\right] \begin{bmatrix} x_{11} \\ x_{21} \\ x_{31} \\ x_{12} \\ x_{22} \\ x_{32} \\ x_{13} \\ x_{23} \\ x_{33} \end{bmatrix}. \tag{4.7}$$

The matrix in (4.7) has a block Toeplitz structure (as indicated by the lines), and each block is itself a Toeplitz matrix. We call such a matrix block Toeplitz with Toeplitz blocks (BTTB). Similar block-structured matrices that arise in image deblurring include

BCCB: Block circulant with circulant blocks;

BHHB: Block Hankel with Hankel blocks;

BTHB: Block Toeplitz with Hankel blocks;

BHTB: Block Hankel with Toeplitz blocks.

With this notation, we can precisely describe the structure of the coefficient matrix $\mathbf{A}$ for the various boundary conditions.

- Zero Boundary Conditions. In this case, $\mathbf{A}$ is a BTTB matrix as demonstrated above.
- Periodic Boundary Conditions. In this case, $\mathbf{A}$ is a BCCB matrix (we will discuss this in Section 4.2).
- Reflexive Boundary Conditions. In this case, $\mathbf{A}$ is a sum of BTTB, BTHB, BHTB, and BHHB matrices (as explained in Section 4.3). Using the notation from Section 3.5, each of these matrices takes into account contributions from $\mathbf{X}$, $\mathbf{X}_{\text{lr}}$, $\mathbf{X}_{\text{ud}}$, and $\mathbf{X}_{\times}$, respectively.

4.1.3 Separable Two-Dimensional Blurs

In some cases the horizontal and vertical components of the blur can be separated, as in our example in Section 1.2. If this is the case, then the $m \times n$ PSF array $\mathbf{P}$ can be decomposed as

$$\mathbf{P} = \mathbf{c}\,\mathbf{r}^T = \begin{bmatrix} c_1 \\ c_2 \\ \vdots \\ c_m \end{bmatrix} \begin{bmatrix} r_1 & r_2 & \cdots & r_n \end{bmatrix},$$

where $\mathbf{r}$ represents the horizontal component of the blur (i.e., blur across the rows of the image array), and $\mathbf{c}$ represents the vertical component (i.e., blur across the columns of the image).

The special structure for this blur implies that $\mathbf{P}$ is a rank-one matrix with elements given by $p_{ij} = c_i\, r_j$. If we insert this relation into the expression in (4.7), we see that the coefficient matrix takes the form

$$\mathbf{A} = \left[\begin{array}{ccc|ccc|ccc}
c_2r_2 & c_1r_2 & & c_2r_1 & c_1r_1 & & & & \\
c_3r_2 & c_2r_2 & c_1r_2 & c_3r_1 & c_2r_1 & c_1r_1 & & & \\
 & c_3r_2 & c_2r_2 & & c_3r_1 & c_2r_1 & & & \\
\hline
c_2r_3 & c_1r_3 & & c_2r_2 & c_1r_2 & & c_2r_1 & c_1r_1 & \\
c_3r_3 & c_2r_3 & c_1r_3 & c_3r_2 & c_2r_2 & c_1r_2 & c_3r_1 & c_2r_1 & c_1r_1 \\
 & c_3r_3 & c_2r_3 & & c_3r_2 & c_2r_2 & & c_3r_1 & c_2r_1 \\
\hline
 & & & c_2r_3 & c_1r_3 & & c_2r_2 & c_1r_2 & \\
 & & & c_3r_3 & c_2r_3 & c_1r_3 & c_3r_2 & c_2r_2 & c_1r_2 \\
 & & & & c_3r_3 & c_2r_3 & & c_3r_2 & c_2r_2
\end{array}\right]$$

$$
= \left[\begin{array}{c|c|c}
r_2 \begin{bmatrix} c_2 & c_1 & \\ c_3 & c_2 & c_1 \\ & c_3 & c_2 \end{bmatrix} & r_1 \begin{bmatrix} c_2 & c_1 & \\ c_3 & c_2 & c_1 \\ & c_3 & c_2 \end{bmatrix} & 0 \\
\hline
r_3 \begin{bmatrix} c_2 & c_1 & \\ c_3 & c_2 & c_1 \\ & c_3 & c_2 \end{bmatrix} & r_2 \begin{bmatrix} c_2 & c_1 & \\ c_3 & c_2 & c_1 \\ & c_3 & c_2 \end{bmatrix} & r_1 \begin{bmatrix} c_2 & c_1 & \\ c_3 & c_2 & c_1 \\ & c_3 & c_2 \end{bmatrix} \\
\hline
0 & r_3 \begin{bmatrix} c_2 & c_1 & \\ c_3 & c_2 & c_1 \\ & c_3 & c_2 \end{bmatrix} & r_2 \begin{bmatrix} c_2 & c_1 & \\ c_3 & c_2 & c_1 \\ & c_3 & c_2 \end{bmatrix}
\end{array}\right].
$$

In general—also for other boundary conditions—the coefficient matrix $\mathbf{A}$ for separable blur has block structure of the form

$$
\mathbf{A} = \mathbf{A}_\mathrm{r} \otimes \mathbf{A}_\mathrm{c} = \begin{bmatrix} a_{11}^{(r)}\mathbf{A}_\mathrm{c} & a_{12}^{(r)}\mathbf{A}_\mathrm{c} & \cdots & a_{1n}^{(r)}\mathbf{A}_\mathrm{c} \\ a_{21}^{(r)}\mathbf{A}_\mathrm{c} & a_{22}^{(r)}\mathbf{A}_\mathrm{c} & \cdots & a_{2n}^{(r)}\mathbf{A}_\mathrm{c} \\ \vdots & \vdots & & \vdots \\ a_{n1}^{(r)}\mathbf{A}_\mathrm{c} & a_{n2}^{(r)}\mathbf{A}_\mathrm{c} & \cdots & a_{nn}^{(r)}\mathbf{A}_\mathrm{c} \end{bmatrix},
$$

where $\mathbf{A}_\mathrm{c}$ is an $m \times m$ matrix, and $\mathbf{A}_\mathrm{r}$ is an $n \times n$ matrix with entries denoted by $a_{ij}^{(r)}$. This special structure, and the symbol $\otimes$ that defines the operation that combines $\mathbf{A}_\mathrm{r}$ and $\mathbf{A}_\mathrm{c}$ in this way, is called a Kronecker product. As we shall see in Section 4.4, the result is that the blurred image $\mathbf{B}$ can be obtained by first convolving each column of $\mathbf{X}$ with $\mathbf{c}$ and then convolving each of the resulting rows with $\mathbf{r}$.

The matrices $\mathbf{A}_\mathrm{r}$ and $\mathbf{A}_\mathrm{c}$ have parameters $\mathbf{r}$ and $\mathbf{c}$, respectively, and they represent the one-dimensional convolutions with the rows and columns. Hence they inherit the structures described in Section 4.1.1. In particular:

- Zero Boundary Conditions. In this case, by (4.4), $\mathbf{A}_\mathrm{r}$ is a Toeplitz matrix with parameters $\mathbf{r}$, and $\mathbf{A}_\mathrm{c}$ is a Toeplitz matrix with parameters $\mathbf{c}$.
- Periodic Boundary Conditions. In this case, using (4.5), $\mathbf{A}_\mathrm{r}$ is a circulant matrix with parameters $\mathbf{r}$, and $\mathbf{A}_\mathrm{c}$ is a circulant matrix with parameters $\mathbf{c}$.
- Reflexive Boundary Conditions. In this case, using (4.6), $\mathbf{A}_\mathrm{r}$ is a Toeplitz-plus-Hankel matrix with parameters $\mathbf{r}$, and $\mathbf{A}_\mathrm{c}$ is a Toeplitz-plus-Hankel matrix with parameters $\mathbf{c}$.

POINTER. Here we list some of the most important properties of Kronecker products; a more extensive list can be found, for example, in [59].

$$
(\mathbf{A}_\mathrm{r} \otimes \mathbf{A}_\mathrm{c})\,\mathrm{vec}(\mathbf{X}) = \mathrm{vec}(\mathbf{A}_\mathrm{c}\,\mathbf{X}\,\mathbf{A}_\mathrm{r}^T),
$$

$$
(\mathbf{A}_\mathrm{r} \otimes \mathbf{A}_\mathrm{c})^T = \mathbf{A}_\mathrm{r}^T \otimes \mathbf{A}_\mathrm{c}^T, \qquad (\mathbf{A}_\mathrm{r} \otimes \mathbf{A}_\mathrm{c})^{-1} = \mathbf{A}_\mathrm{r}^{-1} \otimes \mathbf{A}_\mathrm{c}^{-1},
$$

$$
(\mathbf{U}_\mathrm{r}\boldsymbol{\Sigma}_\mathrm{r}\mathbf{V}_\mathrm{r}^T) \otimes (\mathbf{U}_\mathrm{c}\boldsymbol{\Sigma}_\mathrm{c}\mathbf{V}_\mathrm{c}^T) = (\mathbf{U}_\mathrm{r} \otimes \mathbf{U}_\mathrm{c})(\boldsymbol{\Sigma}_\mathrm{r} \otimes \boldsymbol{\Sigma}_\mathrm{c})(\mathbf{V}_\mathrm{r} \otimes \mathbf{V}_\mathrm{c})^T.
$$

A summary of the important matrix structures described in this section is given in VIP 9. In the following sections we see that it is possible to efficiently compute the SVD or spectral decompositions of some of these matrices. In particular, we see that the matrix $\mathbf{A}$ does not need to be constructed explicitly. Instead we need only work directly with the PSF array.

VIP 9. The following table summarizes the important structures of the blurring matrix, $\mathbf{A}$, described in this section, and how they correspond to the type of PSF (separable or nonseparable) and the type of imposed boundary conditions (zero, periodic, or reflexive).

Boundary condition	Nonseparable PSF	Separable PSF
Zero	BTTB	Kronecker product of Toeplitz matrices
Periodic	BCCB	Kronecker product of circulant matrices
Reflexive	BTTB + BTHB + BHTB + BHHB	Kronecker product of Toeplitz-plus-Hankel matrices

4.2 BCCB Matrices

In this section we consider BCCB matrices that arise in spatially invariant image deblurring when periodic boundary conditions are employed. As an illustration, consider again a 3×3 image. Assuming periodic boundary conditions, element b_{21}, for example, is given by

$$\begin{aligned} b_{21} = \; & p_{33} \cdot x_{13} + p_{32} \cdot x_{11} + p_{31} \cdot x_{12} \\ & + p_{23} \cdot x_{23} + p_{22} \cdot x_{21} + p_{21} \cdot x_{22} \\ & + p_{13} \cdot x_{33} + p_{12} \cdot x_{31} + p_{11} \cdot x_{32}. \end{aligned}$$

It then follows that the linear model now takes the form

$$\begin{bmatrix} b_{11} \\ b_{21} \\ b_{31} \\ b_{12} \\ b_{22} \\ b_{32} \\ b_{13} \\ b_{23} \\ b_{33} \end{bmatrix} = \left[\begin{array}{ccc|ccc|ccc} p_{22} & p_{12} & p_{32} & p_{21} & p_{11} & p_{31} & p_{23} & p_{13} & p_{33} \\ p_{32} & p_{22} & p_{12} & p_{31} & p_{21} & p_{11} & p_{33} & p_{23} & p_{13} \\ p_{12} & p_{32} & p_{22} & p_{11} & p_{31} & p_{21} & p_{13} & p_{33} & p_{23} \\ \hline p_{23} & p_{13} & p_{33} & p_{22} & p_{12} & p_{32} & p_{21} & p_{11} & p_{31} \\ p_{33} & p_{23} & p_{13} & p_{32} & p_{22} & p_{12} & p_{31} & p_{21} & p_{11} \\ p_{13} & p_{33} & p_{23} & p_{12} & p_{32} & p_{22} & p_{11} & p_{31} & p_{21} \\ \hline p_{21} & p_{11} & p_{31} & p_{23} & p_{13} & p_{33} & p_{22} & p_{12} & p_{32} \\ p_{31} & p_{21} & p_{11} & p_{33} & p_{23} & p_{13} & p_{32} & p_{22} & p_{12} \\ p_{11} & p_{31} & p_{21} & p_{13} & p_{33} & p_{23} & p_{12} & p_{32} & p_{22} \end{array}\right] \begin{bmatrix} x_{11} \\ x_{21} \\ x_{31} \\ x_{12} \\ x_{22} \\ x_{32} \\ x_{13} \\ x_{23} \\ x_{33} \end{bmatrix}, \qquad (4.8)$$

and the matrix is BCCB.

4.2.1 Spectral Decomposition of a BCCB Matrix

It is not difficult to show that any BCCB matrix is normal; that is, $\mathbf{A}^* \mathbf{A} = \mathbf{A} \mathbf{A}^*$. Thus $\mathbf{A}$ has a unitary spectral decomposition. It is well known that any BCCB matrix has the particular spectral decomposition

$$\mathbf{A} = \mathbf{F}^* \boldsymbol{\Lambda} \mathbf{F},$$

where $\mathbf{F}$ is the two-dimensional unitary discrete Fourier transform (DFT) matrix. This matrix has a very convenient property: a divide-and-conquer approach can be used to perform matrix-vector multiplications with $\mathbf{F}$ and $\mathbf{F}^*$, without constructing $\mathbf{F}$ explicitly, using fast Fourier transforms (FFTs). In MATLAB, the function `fft2` can be used for matrix-vector multiplications with $\mathbf{F}$, and `ifft2` can be used for multiplications with $\mathbf{F}^*$:

- Implementations of `fft2` and `ifft2` involve a scaling factor that depends on the problem's dimensions. In particular, `fft2` implements matrix-vector multiplication with $\sqrt{N}\mathbf{F}$, and `ifft2` implements multiplication with $\frac{1}{\sqrt{N}}\mathbf{F}^*$, where N is the dimension of $\mathbf{A}$ (i.e., $N = mn$, where the images are $m \times n$ arrays of pixel values). However, the scaling usually takes care of itself since an operation such as `ifft2(fft2(·))` is an identity operation, just as is $\mathbf{F}^*(\mathbf{F}(\cdot))$.

- `fft2` and `ifft2` act on two-dimensional arrays. Thus, to form the multiplication $\mathbf{Fx}$, the vector $\mathbf{x}$ must first be reshaped. For example, if

$$\mathbf{X} = \begin{bmatrix} x_{11} & x_{12} & x_{13} \\ x_{21} & x_{22} & x_{23} \\ x_{31} & x_{32} & x_{33} \end{bmatrix} \quad \longleftrightarrow \quad \mathbf{x} = \mathrm{vec}(\mathbf{X}) = \begin{bmatrix} x_{11} \\ x_{21} \\ x_{31} \\ x_{12} \\ x_{22} \\ x_{32} \\ x_{13} \\ x_{23} \\ x_{33} \end{bmatrix},$$

then

$$\texttt{fft2(X)} \quad \longleftrightarrow \quad \sqrt{N}\mathbf{Fx}.$$

Although this may seem a bit unnatural from a linear algebraic point of view, it is very natural for images. We can perform the product of the matrix with the image data $\mathbf{X}$, rather than the stacked vector representation given by $\mathbf{x}$.

- FFT algorithms are fairly complicated, so it is difficult to give a precise cost for the computations of `fft2(X)` and `ifft2(X)`. In general the speed depends on the size of `X`; they are most efficient if the dimensions have only small prime factors. In particular, if $N = mn = 2^k$, the cost is $O(N \log_2 N)$.

POINTER. If $\mathbf{x}$ is a vector of length n with components x_j, then the (unitary) DFT of $\mathbf{x}$ is the vector $\hat{\mathbf{x}}$ whose kth component is

$$\hat{x}_k = \frac{1}{\sqrt{n}} \sum_{j=1}^{n} x_j e^{-2\pi \hat{\imath}(j-1)(k-1)/n},$$

where $\hat{\imath} = \sqrt{-1}$. The inverse DFT of the vector $\hat{\mathbf{x}}$ is the vector $\mathbf{x}$, where

$$x_j = \frac{1}{\sqrt{n}} \sum_{k=1}^{n} \hat{x}_k e^{2\pi \hat{\imath}(j-1)(k-1)/n}.$$

A two-dimensional DFT of an array $\mathbf{X}$ can be obtained by computing the DFT of the columns of $\mathbf{X}$, followed by a DFT of the rows. A similar approach is used for the inverse two-dimensional DFT.

POINTER. Observe that the DFT and inverse DFT computations can be written as matrix-vector multiplication operations. An FFT is an efficient algorithm to compute this matrix-vector multiplication. Good references on the FFT and its properties include Davis [8], Jain [31], and Van Loan [59], as well as MATLAB's documentation on `fft2` and `ifft2`. See also `http://www.fftw.org`.

We thus have the unitary matrix, but how can we compute the eigenvalues? It turns out that the first column of $\mathbf{F}$ is the vector of all ones, scaled by the square root of the dimension. Denote the first column of $\mathbf{A}$ by $\mathbf{a}_1$ and the first column of $\mathbf{F}$ by $\mathbf{f}_1$ and notice that

$$\mathbf{A} = \mathbf{F}^* \mathbf{\Lambda} \mathbf{F} \quad \Rightarrow \quad \mathbf{F}\mathbf{A} = \mathbf{\Lambda}\mathbf{F} \quad \Rightarrow \quad \mathbf{F}\mathbf{a}_1 = \mathbf{\Lambda}\mathbf{f}_1 = \frac{1}{\sqrt{N}}\boldsymbol{\lambda},$$

where $\boldsymbol{\lambda}$ is a vector containing the eigenvalues of $\mathbf{A}$. Thus, to compute the eigenvalues of $\mathbf{A}$, we need only multiply the matrix $\sqrt{N}\mathbf{F}$ by the first column of $\mathbf{A}$. That is, the eigenvalues can be computed by applying `fft2` to an array containing the first column of $\mathbf{A}$. In our 3×3 example, given explicitly by (4.8), the resulting array is

$$\begin{bmatrix} p_{22} & p_{23} & p_{21} \\ p_{32} & p_{33} & p_{31} \\ p_{12} & p_{13} & p_{11} \end{bmatrix}.$$

Note that this array can be obtained by swapping certain quadrants of entries in the PSF array:

$$\left[\begin{array}{c|cc} p_{11} & p_{12} & p_{13} \\ \hline p_{21} & p_{22} & p_{23} \\ p_{31} & p_{32} & p_{33} \end{array}\right] \quad \longleftrightarrow \quad \left[\begin{array}{cc|c} p_{22} & p_{23} & p_{21} \\ p_{32} & p_{33} & p_{31} \\ \hline p_{12} & p_{13} & p_{11} \end{array}\right]$$

The built-in MATLAB function `circshift` can be used for this purpose. In particular, if `center` is a 1×2 array containing the row and column indices of the center of the PSF array, `P` (in this case, `center = [2, 2]`), then

```
circshift(P, 1 - center);
```

performs the necessary shift. Thus, the spectrum of a BCCB **A**, defined by a PSF array, can be computed very efficiently, without explicitly forming **A**, using the MATLAB statement

```
S = fft2( circshift(P, 1 - center) );
```

If the PSF array is $m \times n$ (and hence **A** is $N \times N$, $N = mn$), then using the FFT to compute the eigenvalues is extremely fast compared to standard approaches (such as MATLAB's `eig` function) which would cost $O(N^3)$ operations.

Now we know that a spectral decomposition of any BCCB matrix is efficient with respect to both time and storage, since the matrix **F** is not computed or stored.

4.2.2 Computations with BCCB Matrices

Once we know the spectral decomposition, we can efficiently perform many matrix computations with BCCB matrices. For example,

$$\mathbf{b} = \mathbf{A}\mathbf{x} = \mathbf{F}^* \mathbf{\Lambda} \mathbf{F} \mathbf{x}$$

can be computed using the MATLAB statements

```
S = fft2( circshift(P, 1 - center) );
B = ifft2( S .* fft2(X) );
B = real(B);
```

The last statement, `B = real(B)`, is used because `fft2` and `ifft2` involve computations with complex numbers. Since the entries in the PSF and **X** are real, the result **B** should contain only real entries. However, due to roundoff errors, the computed **B** may contain small complex parts, which are removed by the statement `B = real(B)`.

Suppose we want to solve the linear system $\mathbf{b} = \mathbf{A}\mathbf{x}$, assuming $\mathbf{A}^{-1}$ exists. Then it is easy to verify that $\mathbf{A}^{-1} = \mathbf{F}^* \mathbf{\Lambda}^{-1} \mathbf{F}$ (by checking that $\mathbf{A}\mathbf{A}^{-1} = \mathbf{I}$), so we simply write

$$\mathbf{x} = \mathbf{A}^{-1}\mathbf{b} = \mathbf{F}^* \mathbf{\Lambda}^{-1} \mathbf{F} \mathbf{b}.$$

Thus, **x** could be computed using the statements

```
S = fft2( circshift(P, 1 - center) );
X = ifft2( fft2(B) ./ S );
X = real(X);
```

It is important to emphasize that because **A** is severely ill-conditioned, the naïve solution, $\mathbf{x}_{\text{naïve}} = \mathbf{A}^{-1}\mathbf{b}$ is likely to be a very poor approximation of the true solution, especially if **b** (i.e., **B**) contains noise (recall the simple examples in Chapter 1). Filtering methods that use the FFT will be discussed in Chapter 5.

VIP 10. When using periodic boundary conditions, basic computations with **A** can be performed using **P**, without ever constructing **A**.

- Given:

  `P` = PSF array
  `center = [row, col]` = center of PSF
  `X` = true image
  `B` = blurred image

- To compute eigenvalues of **A**, use

```
S = fft2( circshift(P, 1 - center) );
```

- To compute the blurred image from the true image, use

```
B = real( ifft2( S .* fft2(X) ) );
```

- To compute the naïve solution from the blurred image, use

```
X = real( ifft2( fft2(B) ./ S ) );
```

4.3 BTTB + BTHB + BHTB + BHHB Matrices

Figure 4.1. *An artificially created PSF that satisfies the double symmetry condition in* (4.10).

In this section we consider the case of using reflexive boundary conditions, so that **A** is a sum of BTTB, BTHB, BHTB, and BHHB matrices.

This seems fairly complicated, but the matrix has a simple structure if the nonzero part of the PSF satisfies a double symmetry condition. In particular, suppose that the $m \times n$ PSF array, $\mathbf{P}$, has the form

$$\mathbf{P} = \begin{bmatrix} \mathbf{0} & \mathbf{0} & \mathbf{0} \\ \mathbf{0} & \widetilde{\mathbf{P}} & \mathbf{0} \\ \mathbf{0} & \mathbf{0} & \mathbf{0} \end{bmatrix}, \tag{4.9}$$

where $\widetilde{\mathbf{P}}$ is $(2k-1) \times (2k-1)$ with center located at the (k, k) entry, and where the zero blocks may have different (but consistent) dimensions. If

$$\widetilde{\mathbf{P}} = \texttt{fliplr}(\widetilde{\mathbf{P}}) = \texttt{flipud}(\widetilde{\mathbf{P}}), \tag{4.10}$$

where `fliplr` and `flipud` are MATLAB functions that flip the columns (rows) of an array in a left-right (up-down) direction, then we say that the PSF satisfies a double symmetry condition. For example, the arrays

$$\begin{bmatrix} 1 & 2 & 1 \\ 3 & 4 & 3 \\ 1 & 2 & 1 \end{bmatrix}, \quad \begin{bmatrix} 1 & 2 & 1 & 0 \\ 3 & 4 & 3 & 0 \\ 1 & 2 & 1 & 0 \\ 0 & 0 & 0 & 0 \end{bmatrix}, \quad \begin{bmatrix} 0 & 0 & 0 & 0 \\ 0 & 1 & 2 & 1 \\ 0 & 3 & 4 & 3 \\ 0 & 1 & 2 & 1 \\ 0 & 0 & 0 & 0 \end{bmatrix}$$

are doubly symmetric. Figure 4.1 shows an example of a doubly symmetric PSF.

This symmetry condition does occur in practice; for example, the Gaussian model for atmospheric turbulence blur. The result is that the matrix $\mathbf{A}$ is symmetric. In addition, the matrix is block symmetric, and each block is itself symmetric. In this case, it can be shown that $\mathbf{A}$ is normal, and that it has the real spectral decomposition

$$\mathbf{A} = \mathbf{C}^T \mathbf{\Lambda} \mathbf{C},$$

where $\mathbf{C}$ is the orthogonal two-dimensional discrete cosine transformation (DCT) matrix [44].

The matrix $\mathbf{C}$ is very similar to $\mathbf{F}$; it is highly structured, and there are fast algorithms for computing matrix-vector multiplications with $\mathbf{C}$ and $\mathbf{C}^T$. In MATLAB, `dct2` and `idct2` can be used for these computations. The cost and storage requirements are of the same order as those of `fft2` and `ifft2`, but savings can be achieved by the use of real arithmetic rather than complex.

POINTER. If $\mathbf{x}$ is a vector of length n with components x_j, then the (orthogonal) DCT of $\mathbf{x}$ is the vector $\hat{\mathbf{x}}$, where

$$\hat{x}_k = \omega_k \sum_{j=1}^{n} x_j \cos \frac{\pi(2j-1)(k-1)}{2n},$$

where $\omega_1 = \sqrt{1/n}$ and $\omega_k = \sqrt{2/n}$, $k = 2, \ldots, n$. The inverse DCT of the vector $\hat{\mathbf{x}}$ is the vector $\mathbf{x}$, where

$$x_j = \sum_{k=1}^{n} \omega_k \hat{x}_k \cos \frac{\pi(2j-1)(k-1)}{2n},$$

where ω_k is defined above. A two-dimensional DCT of an array $\mathbf{X}$ can be obtained by computing the DCT of the columns of $\mathbf{X}$, followed by a DCT of the rows. A similar approach is used for the inverse two-dimensional DCT.

As with the BCCB case, the eigenvalues are easily computed:

$$\lambda_i = \frac{[\mathbf{C}\mathbf{a}_1]_i}{c_{i1}},$$

where $\mathbf{a}_1$ is the first column of $\mathbf{A}$ and c_{i1} is an element of $\mathbf{C}$. Thus, to compute the eigenvalues, we need only construct the first column of $\mathbf{A}$ and use the `dct2` function. The first column can be found by basic manipulation of the PSF array. For example, suppose the PSF array is given by

$$\mathbf{P} = \begin{bmatrix} p_{11} & p_{12} & p_{13} & p_{14} & p_{15} \\ p_{21} & p_{22} & p_{23} & p_{24} & p_{25} \\ p_{31} & p_{32} & p_{33} & p_{34} & p_{35} \\ p_{41} & p_{42} & p_{43} & p_{44} & p_{45} \\ p_{51} & p_{52} & p_{53} & p_{54} & p_{55} \end{bmatrix}.$$

If reflexive boundary conditions are used, then the first column of $\mathbf{A}$ (which is a 25×25 BTTB + BTHB + BHTB + BHHB matrix) can be represented by the array

$$\begin{bmatrix} p_{33} & p_{34} & p_{35} & 0 & 0 \\ p_{43} & p_{44} & p_{45} & 0 & 0 \\ p_{53} & p_{54} & p_{55} & 0 & 0 \\ 0 & 0 & 0 & 0 & 0 \\ 0 & 0 & 0 & 0 & 0 \end{bmatrix} + \begin{bmatrix} p_{34} & p_{35} & 0 & 0 & 0 \\ p_{44} & p_{45} & 0 & 0 & 0 \\ p_{54} & p_{55} & 0 & 0 & 0 \\ 0 & 0 & 0 & 0 & 0 \\ 0 & 0 & 0 & 0 & 0 \end{bmatrix}$$

$$+ \begin{bmatrix} p_{43} & p_{44} & p_{45} & 0 & 0 \\ p_{53} & p_{54} & p_{55} & 0 & 0 \\ 0 & 0 & 0 & 0 & 0 \\ 0 & 0 & 0 & 0 & 0 \\ 0 & 0 & 0 & 0 & 0 \end{bmatrix} + \begin{bmatrix} p_{44} & p_{45} & 0 & 0 & 0 \\ p_{54} & p_{55} & 0 & 0 & 0 \\ 0 & 0 & 0 & 0 & 0 \\ 0 & 0 & 0 & 0 & 0 \\ 0 & 0 & 0 & 0 & 0 \end{bmatrix}.$$

POINTER. The MATLAB functions `dct2` and `idct2` are only included with the IPT. If this is not available, it is not too difficult to write similar functions that use `fft2` and `ifft2`; see the functions `dcts`, `idcts`, `dcts2`, and `idcts2` in Appendix 3. The relationship between the DCT and FFT can be found, for example, in Jain [31] and Van Loan [59].

The first term accounts for pixels within the frame of the image, the second for pixels to the left, the third for pixels above, and the last for pixels above and to the left. In general, we can construct an array containing the first column of $\mathbf{A}$ directly from the PSF array as follows:

- Suppose the PSF array, $\mathbf{P}$, is $(2k-1) \times (2k-1)$, with the center located at the (k, k) entry.
- Define the shift matrices $\mathbf{Z}_1$ and $\mathbf{Z}_2$ using the MATLAB functions `diag` and `ones` (illustrated for $k = 3$):

$$\mathbf{Z}_1 = \texttt{diag(ones(k,1), k-1)} = \begin{bmatrix} 0 & 0 & 1 & 0 & 0 \\ 0 & 0 & 0 & 1 & 0 \\ 0 & 0 & 0 & 0 & 1 \\ 0 & 0 & 0 & 0 & 0 \\ 0 & 0 & 0 & 0 & 0 \end{bmatrix},$$

$$\mathbf{Z}_2 = \texttt{diag(ones(k-1,1), k)} = \begin{bmatrix} 0 & 0 & 0 & 1 & 0 \\ 0 & 0 & 0 & 0 & 1 \\ 0 & 0 & 0 & 0 & 0 \\ 0 & 0 & 0 & 0 & 0 \\ 0 & 0 & 0 & 0 & 0 \end{bmatrix}.$$

- Then an array containing the first column of the corresponding blurring matrix can be constructed as

$$\mathbf{Z}_1\mathbf{P}\mathbf{Z}_1^T + \mathbf{Z}_1\mathbf{P}\mathbf{Z}_2^T + \mathbf{Z}_2\mathbf{P}\mathbf{Z}_1^T + \mathbf{Z}_2\mathbf{P}\mathbf{Z}_2^T.$$

It is not difficult to generalize this to the case when $\mathbf{P}$ has arbitrary dimension and center, provided it has the double symmetry structure about the center defined by (4.9) and (4.10). In general, we just need to know the pixel location of the center of the PSF. A MATLAB function implementing this process, `dctshift`, can be found in Appendix 3 (there is no analogous built-in function).

Once the first column of $\mathbf{A}$ is known, it is a simple matter to compute its spectrum using the `dct2` function:

```
e1 = zeros(size(P));
e1(1,1) = 1;
S = dct2( dctshift(P, center) ) ./ dct2( e1 );
```

where we use `e1` to denote the array version of the first unit vector, $\mathbf{e}_1$.

As with BCCB matrices, once we know how to compute the spectral decomposition, we can efficiently perform matrix computations with these doubly symmetric BTTB + BTHB + BHTB + BHHB matrices. The results are summarized in VIP 11.

VIP 11. When using reflexive boundary conditions and a doubly symmetric PSF, basic computations with **A** can be performed using **P**, without ever constructing **A**.

- Given:

  `P` = PSF array
  `center = [row, col]` = center of PSF
  `X` = true image
  `B` = blurred image

- To compute eigenvalues of **A**, use

```
e1 = zeros(size(P));, e1(1,1) = 1;
S = dct2( dctshift(P, center) ) ./ dct2(e1);
```

- To compute the blurred image from the true image, use

```
B = idct2( S .* dct2(X) );
```

- To compute the naïve solution from the blurred image, use

```
X = idct2( dct2(B) ./ S );
```

- If the IPT is not available, then use `dcts2` and `idcts2` in place of `dct2` and `idct2`. Implementations of `dcts2` and `idcts2` may be obtained from the book's website.

4.4 Kronecker Product Matrices

In this section we consider blurring matrices, **A**, that can be represented as a Kronecker product,

$$\mathbf{A} = \mathbf{A}_{\mathrm{r}} \otimes \mathbf{A}_{\mathrm{c}}, \tag{4.11}$$

where, if the images **X** and **B** have $m \times n$ pixels, then $\mathbf{A}_{\mathrm{r}}$ is $n \times n$ and $\mathbf{A}_{\mathrm{c}}$ is $m \times m$. In Section 4.4.1 we show that it is possible to recognize if such a representation exists, and how to construct the smaller matrices $\mathbf{A}_{\mathrm{r}}$ and $\mathbf{A}_{\mathrm{c}}$ directly from the PSF array. We then discuss how to efficiently implement important matrix computations with Kronecker products in Section 4.4.2.

4.4.1 Constructing the Kronecker Product from the PSF

Suppose the PSF array, **P**, is an $m \times n$ array, and that it can be represented as an outer product of two vectors (i.e., the PSF is separable),

$$\mathbf{P} = \mathbf{c}\,\mathbf{r}^T = \begin{bmatrix} c_1 \\ c_2 \\ \vdots \\ c_m \end{bmatrix} \begin{bmatrix} r_1 & r_2 & \cdots & r_n \end{bmatrix}.$$

Then, as shown in Section 4.1.3, the blurring matrix, **A**, can be represented as the Kronecker product (4.11) using the matrices $\mathbf{A}_c$ and $\mathbf{A}_r$ defined in Section 1.2. The structures of $\mathbf{A}_r$ and $\mathbf{A}_c$ depend on the imposed boundary condition. For example, if we use a 3×3 PSF array, with center (2, 2), then

$$\mathbf{P} = \begin{bmatrix} c_1 \\ c_2 \\ c_3 \end{bmatrix} \begin{bmatrix} r_1 & r_2 & r_3 \end{bmatrix} = \begin{bmatrix} c_1r_1 & c_1r_2 & c_1r_3 \\ c_2r_1 & c_2r_2 & c_2r_3 \\ c_3r_1 & c_3r_2 & c_3r_3 \end{bmatrix}.$$

Then, for zero boundary conditions, $\mathbf{A} = \mathbf{A}_r \otimes \mathbf{A}_c$, where

$$\mathbf{A}_r = \begin{bmatrix} r_2 & r_1 & \\ r_3 & r_2 & r_1 \\ & r_3 & r_2 \end{bmatrix} \quad \text{and} \quad \mathbf{A}_c = \begin{bmatrix} c_2 & c_1 & \\ c_3 & c_2 & c_1 \\ & c_3 & c_2 \end{bmatrix}.$$

To explicitly construct $\mathbf{A}_r$ and $\mathbf{A}_c$ from the PSF array, **P**, we need to be able to find the vectors **r** and **c**. This can be done by computing the largest singular value, and corresponding singular vectors, of **P**. In MATLAB, we can do this efficiently with the built-in `svds` function:

```
[u, s, v] = svds(P, 1);
c = sqrt(s)*u;
r = sqrt(s)*v;
```

In a careful implementation we would compute the first two singular values and check that the ratio of s_2/s_1 is small enough to neglect all but the first and hence that the Kronecker product representation $\mathbf{A}_r \otimes \mathbf{A}_c$ is an accurate representation of **A**. For details, see the function `kronDecomp` in Appendix 3.

4.4.2 Matrix Computations with Kronecker Products

Kronecker products have many important properties that can be exploited when performing matrix computations. For example, if $\mathbf{A}_r$ is $n \times n$ and $\mathbf{A}_c$ is $m \times m$, then $\mathbf{A} = \mathbf{A}_r \otimes \mathbf{A}_c$ is $mn \times mn$, and the matrix-vector relation

$$\mathbf{b} = \mathbf{A}\mathbf{x} \tag{4.12}$$

is equivalent to the matrix-matrix relation

$$\mathbf{B} = \mathbf{A}_c\mathbf{X}\mathbf{A}_r^T, \tag{4.13}$$

exploited in Section 1.2, where $\mathbf{x} = \text{vec}(\mathbf{X})$ and $\mathbf{b} = \text{vec}(\mathbf{B})$. Because the dimensions of $\mathbf{A}_r$ and $\mathbf{A}_c$ are relatively small (on the order of the image pixel dimensions), it is possible to construct the matrices explicitly. In addition, matrix-matrix computations are trivial (and efficient) to implement in MATLAB:

```
B = Ac*X*Ar';
```

Similarly, if $\mathbf{A}_r$ and $\mathbf{A}_c$ are nonsingular, then the solution of $(\mathbf{A}_r \otimes \mathbf{A}_c)\mathbf{x} = \mathbf{b}$ can be represented as

$$\mathbf{X} = \mathbf{A}_c^{-1}\mathbf{B}\mathbf{A}_r^{-T}$$

and easily computed in MATLAB using the backslash and forward slash operators:

```
X = Ac \ B / Ar';
```

We remark that if $\mathbf{A}_\text{r} = \mathbf{A}_\text{c}$, then it is more efficient to compute an LU factorization of $\mathbf{A}_\text{r}$ and compute

```
X = ((U \ (L \ B)) / L') / U';
```

The SVD of a Kronecker product can be expressed in terms of the SVDs of the matrices that form it; if

$$\mathbf{A}_\text{r} = \mathbf{U}_\text{r}\mathbf{\Sigma}_\text{r}\mathbf{V}_\text{r}^T \quad \text{and} \quad \mathbf{A}_\text{c} = \mathbf{U}_\text{c}\mathbf{\Sigma}_\text{c}\mathbf{V}_\text{c}^T,$$

then

$$\mathbf{A} = \mathbf{A}_\text{r} \otimes \mathbf{A}_\text{c} = (\mathbf{U}_\text{r}\mathbf{\Sigma}_\text{r}\mathbf{V}_\text{r}^T) \otimes (\mathbf{U}_\text{c}\mathbf{\Sigma}_\text{c}\mathbf{V}_\text{c}^T) = (\mathbf{U}_\text{r} \otimes \mathbf{U}_\text{c})(\mathbf{\Sigma}_\text{r} \otimes \mathbf{\Sigma}_\text{c})(\mathbf{V}_\text{r} \otimes \mathbf{V}_\text{c})^T.$$

The final matrix factorization is essentially the SVD of $\mathbf{A}$, except that it does not satisfy the usual requirement that the diagonal entries of $\mathbf{\Sigma}_\text{r} \otimes \mathbf{\Sigma}_\text{c}$ appear in nonincreasing order.

Note that we do not need to explicitly form the big matrices $\mathbf{U}_\text{r} \otimes \mathbf{U}_\text{c}$, etc. Instead we work directly with the small matrices $\mathbf{U}_\text{r}$, $\mathbf{U}_\text{c}$, etc. For example, the naïve solution computed using the SVD,

$$\mathbf{x}_\text{naïve} = \mathbf{A}^{-1}\mathbf{b} = \mathbf{V}\mathbf{\Sigma}^{-1}\mathbf{U}^T\mathbf{b},$$

can be written as

$$\mathbf{X}_\text{naïve} = \mathbf{A}_\text{c}^{-1}\mathbf{B}\mathbf{A}_\text{r}^{-T} = \mathbf{V}_\text{c}\mathbf{\Sigma}_\text{c}^{-1}\mathbf{U}_\text{c}^T\mathbf{B}\mathbf{U}_\text{r}\mathbf{\Sigma}_\text{r}^{-1}\mathbf{V}_\text{r}^T.$$

Notice that if $\widehat{\mathbf{B}} = \mathbf{U}_\text{c}^T\mathbf{B}\mathbf{U}_\text{r}$, then, using the MATLAB ./ notation for elementwise division,

$$\mathbf{\Sigma}_\text{c}^{-1}\widehat{\mathbf{B}}\mathbf{\Sigma}_\text{r}^{-1} = \widehat{\mathbf{B}}./\mathbf{S},$$

where $\mathbf{S}$ is an $m \times n$ array containing the singular values of $\mathbf{A}$, arranged such that vec($\mathbf{S}$) holds the diagonal elements of $\mathbf{\Sigma}_r \otimes \mathbf{\Sigma}_c$. Thus, $\mathbf{X}_\text{naïve}$ could be computed in MATLAB as

```
[Uc, Sc, Vc] = svd(Ac);
[Ur, Sr, Vr] = svd(Ar);
S = diag(Sc) * diag(Sr)';
X = Vc * ( (Uc' * B * Ur)./S ) * Vr';
```

where `S = diag(Sc) * diag(Sr)'` is an efficient way to compute the array of singular values. (MATLAB's `diag` function extracts the diagonal elements of a matrix and puts them into a vector; see `help diag` for more information.)

We summarize the above computations in the following VIP. Note that if the PSF is separable, then computations can be implemented efficiently for any boundary condition, and we do not have the restrictions imposed for the FFT- and DCT-based methods.

VIP 12. When using a separable PSF, basic computations with **A** can be performed using **P**, without ever constructing **A**.

- Given:

  `P` = PSF array
  `center = [row, col]` = center of PSF
  `X` = true image
  `B` = blurred image
  `BC` = string denoting boundary condition (e.g., `'zero'`)

- To construct the Kronecker product terms, $\mathbf{A}_r$ and $\mathbf{A}_c$, use

```
[Ar, Ac] = kronDecomp(P, center, BC);
```

  For details, see the function `kronDecomp` in Appendix 3.

- To compute singular values of **A**, use

```
sr = svd(Ar);, sc = svd(Ac);
S = sc * sr';
```

- To compute the blurred image from the true image, use

```
B = Ac * X * Ar';
```

- To compute the naïve solution from the blurred image, use

```
X = Ac \ B / Ar';
```

- To use the SVD to compute the naïve solution from the blurred image, use

```
[Uc, Sc, Vc] = svd(Ac);
[Ur, Sr, Vr] = svd(Ar);
S = diag(Sc) * diag(Sr)';
X = Vc * ( (Uc' * B * Ur)./S ) * Vr';
```

4.5 Summary of Fast Algorithms

In the previous three sections we have described three fast algorithms for computing a spectral decomposition or an SVD. The three algorithms use certain structures of the matrices, and they are therefore tailored to problems with specific PSFs and specific boundary conditions. We summarize this information in the following VIP.

VIP 13. For spatially invariant PSFs, we have the following fast algorithms (recall the table on matrix structures given in VIP 9).

PSF	Boundary condition	Matrix structure	Fast algorithm
Arbitrary	Periodic	BCCB	Two-dimensional FFT
Doubly symmetric	Reflexive	BTTB + BTHB + BHTB + BHHB	Two-dimensional DCT
Separable	Arbitrary	Kronecker product	2 small SVDs

4.6 Creating Realistic Test Data

Image deblurring test data is often created artificially by blurring a known true image with a known PSF. Random values are then added to the blurred pixel values to simulate additive noise. In order to create such test data using the computational techniques discussed in this section, a few issues need to be considered. The examples and sample MATLAB code in this section require data and certain M-files that can be obtained from the book's website.

It is first necessary to say a few words about the PSF. Recall from Section 3.3 that generally the light intensity of the PSF is confined to a small area around its center, beyond which the intensity is essentially zero. Therefore, to conserve storage, it is often the case that the PSF array, **P**, has much smaller dimensions than the associated image arrays, **X** and **B**. For example, the Gaussian PSF arrays

```
Psmall = psfGauss([31, 31], 4);
Pbig = psfGauss([63, 63], 4);
```

essentially contain the same information about the PSF. However, if we want to use the computational techniques outlined in this chapter, then **P** must have the same shape as **X** and **B**. It is a simple matter to extend the dimensions of **P** by padding with zeros. For example, if `Psmall` is a given, small PSF array, and we want to pad with zeros so that it has the same dimension as a large image, `Xbig`, then we can simply use the MATLAB statements

```
Pbig = zeros(size(Xbig));
Pbig(1:size(Psmall,1), 1:size(Psmall,2)) = Psmall;
```

Because we need to do this kind of zero padding fairly often, we created a simple function, `padPSF`, for this purpose (see Appendix 3). Specifically, the zero padding can be done as

```
Pbig = padPSF(Psmall, size(Xbig));
```

Note that the zero padding could be done in a variety of ways, but by keeping `Psmall` in the upper left corner of `Pbig`, both PSF arrays have the same center. This is convenient because computations involving the PSF usually require knowledge of its center. We remark that padding trades storage for time, and if storage is an issue, then alternative schemes, such as overlap-add and overlap-save techniques, can be used; see, for example, [54].

We now turn to the issue of creating a blurred image. To perform the blurring operation, we must first choose a matrix model (i.e., fix a boundary condition) for **A**. For example, we could use the approach outlined in VIP 10 to create the blurred image. The problem with this approach is that we must enforce a specific boundary condition on the blurring operation, which may not accurately model the actual scene. Note that if we are *given* blurred image data, then we try to make a best guess at what boundary condition is most realistic, but if we are *creating* blurred image data, then we should try to simulate blurring using correct data from the larger scene as illustrated in Figure 4.2. This can be done by performing the blurring operation on a large image, from which a central part is extracted. Any boundary condition can be used to perform the blurring operation on the large image. Consider, for example, the following MATLAB statements:

```
Xbig = double(imread('iograyBorder.tif'));
[P, center] = psfGauss([512,512], 6);
Pbig = padPSF(P, size(Xbig));
Sbig = fft2(circshift(Pbig, 1-center));
Bbig = real(ifft2(Sbig .* fft2(Xbig)));
X = Xbig(51:562,51:562);
B = Bbig(51:562,51:562);
```

These computations provide test data consisting of a set of 512×512 images: the PSF array, `P`, a "true" image, `X`, and a blurred image `B`. Because the nonzero extent of the PSF is much smaller than the number of rows and columns that were discarded from the "big" images, the image `B` is a realistic representation of a blurred image taken from an infinite scene. Note also that we could take more rows and columns from `Xbig` and `Bbig`, as long as the number of discarded rows and columns around the boundary is at least half the diameter of the nonzero extent of the PSF.

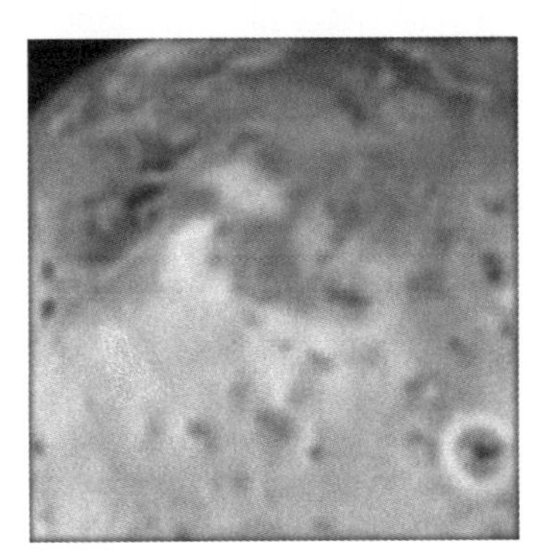

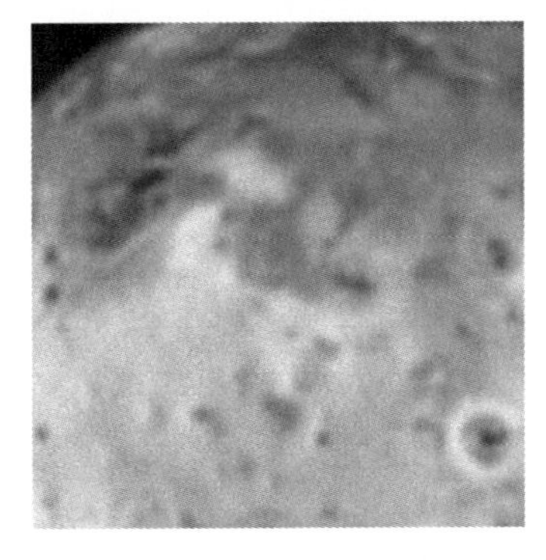

Figure 4.2. *The left image was created using a blurring matrix* **A** *with zero boundary conditions, leading to artificial dark edges. The right image was created via a blurring of a larger image, followed by extraction of the central part.*

The final step in creating realistic test data is to include additive noise. This can be done by adding random perturbations to the entries in the blurred image. For example, using the built-in MATLAB function `randn`, we can add Gaussian white noise so that $\|\mathbf{e}\|_2/\|\mathbf{Ax}\|_2 = 0.01$; that is, we add 1% noise to the blurred data:

```
E = randn(size(B));
E = E / norm(E,'fro');
B = B + 0.01*norm(B,'fro')*E;
```

Notice that before adding the noise in the third step in the above computation, $\mathtt{B} = \text{vec}(\mathbf{Ax})$, the noise-free blurred image. Also observe that `norm(E,'fro')` and `norm(B,'fro')` compute Frobenius norms of arrays `E` and `B`, which is equivalent to computing 2-norms of vectors $\mathbf{e}$ and $\mathbf{Ax}$.

We will use this process to generate test problems in the remainder of the book. The specific example considered in this section is implemented in `chapter4demo.m`, which may be obtained from the book's website.

CHALLENGE 9. For the image `ioborder.gif`, use the PSF for Gaussian blur (with $s_1 = s_2$) or Moffat blur (with $s_1 = s_2$ and $\beta = 3$) together with the bordering approach from Section 4.6 to generate noisy images for different noise levels. Then use the fast algorithm in VIP 11 to compute the naïve solution. For each noise level, how large can you make the parameters $s_1 = s_2$ in the PSF before the errors start to dominate the reconstruction? What happens if you perform the same tests using the fast algorithm in VIP 10?

CHALLENGE 10. For small separable PSFs it is possible to construct the blurring matrix $\mathbf{A}$ explicitly using our `kronDecomp` function and the MATLAB built-in function `kron`. Do this for $n \times n$ Gaussian PSFs with $n = 16, 23, 32, 47$. Use the MATLAB `tic` and `toc` functions to measure the time required to compute the singular values and eigenvalues of $\mathbf{A}$. Compare this with the time required to compute the same quantities (i.e., the array `S`) using the efficient methods given in VIPs 10, 11, and 12. Now use these efficient approaches for larger $n \times n$ Gaussian PSFs, with $n = 64, 81, 101, 128, 181, 243, 256, 349, 512, 613, 729, 887, 1024$. What do these results suggest? For example, can you explain why one approach is faster/slower than another? Can you explain the FFT and DCT timings for $n = 887$ compared to the timings for $n = 729$ and $n = 1024$?

Chapter 5

SVD and Spectral Analysis

Science is spectral analysis. Art is light synthesis.
– Karl Kraus

The previous chapter shows that it is easy to solve noise-free problems that have structure, but what should we do when noise is present? The answer is to filter the solution in order to diminish the effects of noise in the data. In this chapter we use the SVD to analyze and build up an understanding of the mechanisms and difficulties associated with image deblurring.

5.1 Introduction to Spectral Filtering

Recall from Section 1.4 that we define the SVD of the $N \times N$ matrix $\mathbf{A}$ to be

$$\mathbf{A} = \mathbf{U}\,\mathbf{\Sigma}\,\mathbf{V}^T, \tag{5.1}$$

where $\mathbf{U}$ and $\mathbf{V}$ are orthogonal matrices, satisfying $\mathbf{U}^T\mathbf{U} = \mathbf{I}_N$ and $\mathbf{V}^T\mathbf{V} = \mathbf{I}_N$, and $\mathbf{\Sigma}$ is a diagonal matrix with entries $\sigma_1 \geq \sigma_2 \geq \cdots \geq \sigma_N \geq 0$. For a blurring matrix, all the singular values decay gradually to zero and the condition number $\text{cond}(\mathbf{A}) = \sigma_1/\sigma_N$ is very large.

Our example at the end of Chapter 1 introduced the SVD as a tool for analysis in image deblurring. In particular, we used the SVD to explain how the noise $\mathbf{e}$ in the data (the recorded blurred image) enters the reconstructed image in the form of the inverted noise $\mathbf{A}^{-1}\mathbf{e}$.

The SVD analysis that we carried out is independent of the algorithm that we choose to solve the image deblurring problem, but it does suggest the SVD as one method for dealing with the inverted noise. That is, we might try to use the SVD approach to damp the effects caused by division by the small singular values.

Recall that the naïve solution (cf. (1.5)–(1.7)) can be written as

$$\mathbf{x}_{\text{naïve}} = \mathbf{A}^{-1}\mathbf{b} = \mathbf{V}\,\mathbf{\Sigma}^{-1}\mathbf{U}^T\mathbf{b} = \sum_{i=1}^{N} \frac{\mathbf{u}_i^T\mathbf{b}}{\sigma_i}\,\mathbf{v}_i\,.$$

One approach to damp the effects caused by division of small singular values is to simply discard all SVD components that are dominated by noise—typically the ones for indices i above a certain truncation parameter k. The resulting method is, for obvious reasons, referred to as the truncated SVD, or TSVD, method,[6] and it amounts to computing an approximate solution of the form

$$\mathbf{x}_k = \sum_{i=1}^{k} \frac{\mathbf{u}_i^T \mathbf{b}}{\sigma_i} \mathbf{v}_i, \qquad k < N. \tag{5.2}$$

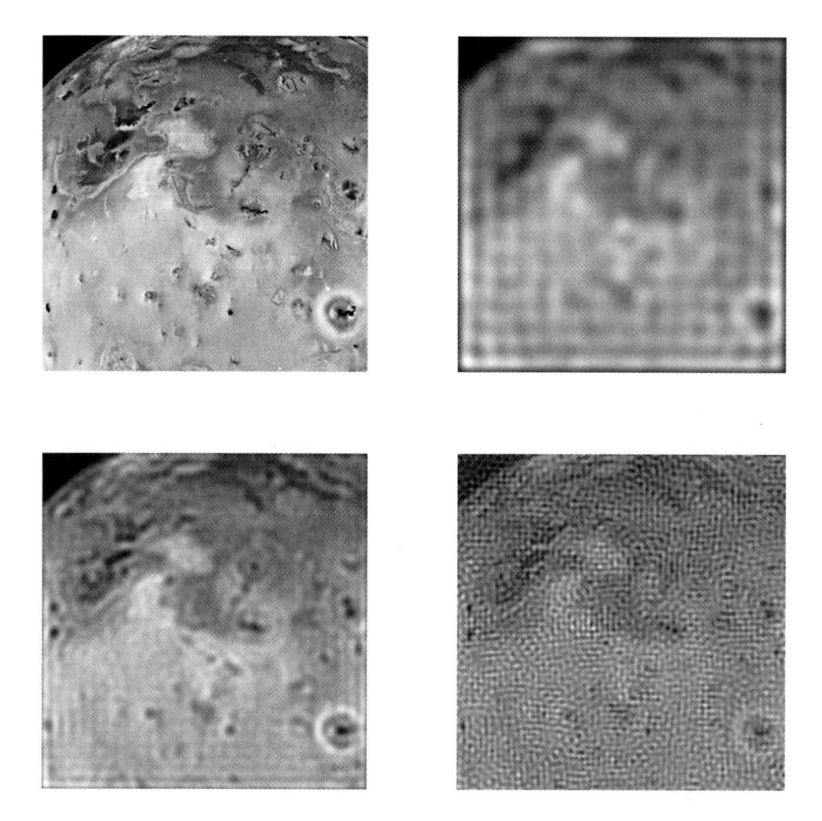

Figure 5.1. *Exact image (top left) and three TSVD solutions $\mathbf{x}_k$ to the image deblurring problem, computed for three different values of the truncation parameter: $k =$ 658 (top right), $k = 2813$ (bottom left), and $k = 7243$ (bottom right). The corresponding solutions range from oversmoothed to undersmoothed, as k goes from small to large values.*

In spite of its simplicity, this method can work quite well. Figure 5.1 shows three TSVD solutions $\mathbf{x}_k$ computed for three different values of the truncation parameter k. The original image is `iogray.tif`, which was blurred with a Gaussian PSF, followed by the addition of Gaussian white noise $\mathbf{e}$ with $\|\mathbf{e}\|_2/\|\mathbf{b}\|_2 = 0.05$.

[6]It is occasionally referred to as the pseudo-inverse filter.

Note that as k increases, more terms are included in the SVD expansion, and consequently components with higher frequencies are included—hence we can think of k as a way to control how much smoothing (or low-pass filtering) is introduced in the reconstruction. The smallest value $k = 658$ was deliberately chosen too small, to show the effect of oversmoothing, while the largest value $k = 7243$ shows an undersmoothed reconstruction with too much influence from the high-frequency components of the noise.

The TSVD method is an example of the general class of methods that are called spectral filtering methods, which have the form

$$\mathbf{x}_{\text{filt}} = \sum_{i=1}^{N} \phi_i \frac{\mathbf{u}_i^T \mathbf{b}}{\sigma_i} \mathbf{v}_i \, , \tag{5.3}$$

where the filter factors ϕ_i are chosen such that $\phi_i \approx 1$ for large singular values, and $\phi_i \approx 0$ for small singular values. Different spectral filtering algorithms involve different choices of the filter factors.

General image deblurring algorithms inevitably involve some kind of filtering in order to damp the influence of the noise. The question is, how should this filtering be done, and how much filtering should be used? A large amount of filtering ensures that noise is highly suppressed, but at the cost of loss of information. Thus we really want to determine a filter that balances the details that can be recovered with the influence of the noise. This chapter describes a class of deblurring algorithms based on filtering methods. We give some examples of commonly used filters and show how to implement them efficiently in MATLAB. There are several important issues that need to be addressed: choosing the filter factors, choosing proper bases (we might prefer to use a Fourier basis instead of the SVD), and designing efficient implementations for large-scale problems.

VIP 14. Spectral filtering amounts to filtering each of the components of the solution in the spectral basis, and in such a way that the influence from the noise in the blurred image is damped.

5.2 Incorporating Boundary Conditions

In the example used in Figure 5.1, we implicitly assumed zero boundary conditions; cf. Section 3.5. This assumption manifests itself as clearly visible oscillations in the reconstructions having their largest amplitude near the borders of the image. Now we illustrate how the use of other boundary conditions can reduce these oscillations and thus improve the quality of the reconstructed image.

As explained in Section 4.6 we first blur a large exact image to produce the large and blurred image, and then we "chop off" the borders of this image to obtain the resulting image $\mathbf{B}$. In this way, any artifacts from boundary conditions in the model problem will appear outside the borders of the chopped image $\mathbf{B}$. Finally, noise is added to this image. The resulting blurred image is shown on the right in Figure 4.2.

The problem that we need to solve in order to compute the desired reconstruction is

$$\mathbf{A}\mathbf{x} = \mathbf{b} \qquad \text{with} \qquad \mathbf{A} = \mathbf{A}_0 + \mathbf{A}_{\text{BC}}, \tag{5.4}$$

where $\mathbf{A}_0$ is the BTTB matrix resulting from zero boundary conditions, and $\mathbf{A}_{\mathrm{BC}}$ is a boundary correction term that incorporates specific boundary conditions into the model. Similar to $\mathbf{A}$, the matrix $\mathbf{A}_{\mathrm{BC}}$ is structured, and (as we saw in Section 4) its form depends on the type of boundary condition. Reconstructions based on $\mathbf{A} = \mathbf{A}_0$ correspond to zero boundary conditions.

Independent of the specified boundary condition, the modified problem (5.4) has the characteristics that make image deblurring very challenging: gradually decaying singular values, a very large condition number, and a severe sensitivity to the noise in the data. Hence, we must also use a spectral filtering method to solve the modified problem. The only change is that we must now use the SVD of the corrected matrix $\mathbf{A}$ (instead of the SVD of $\mathbf{A}_0$).

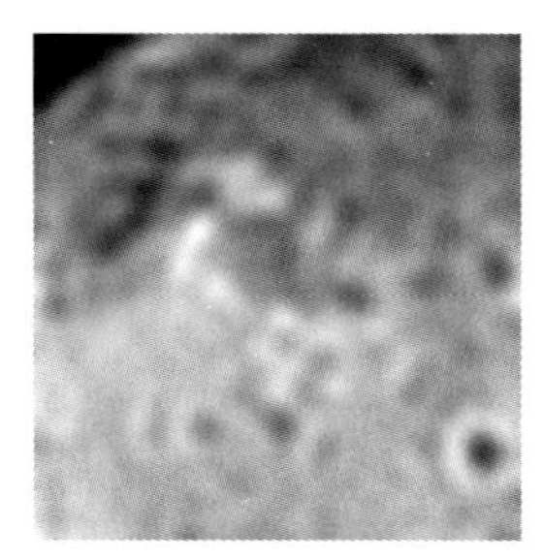 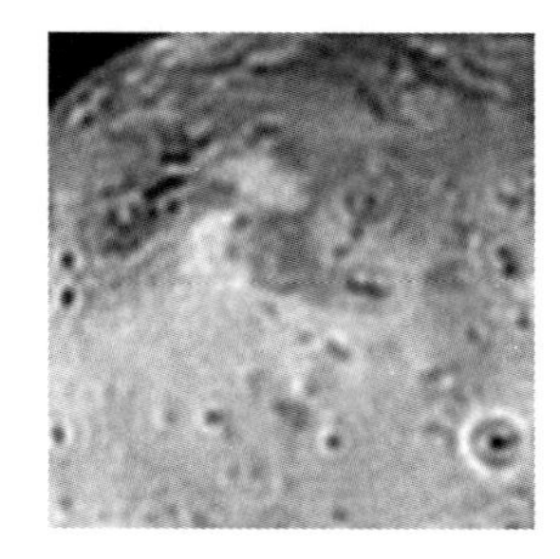 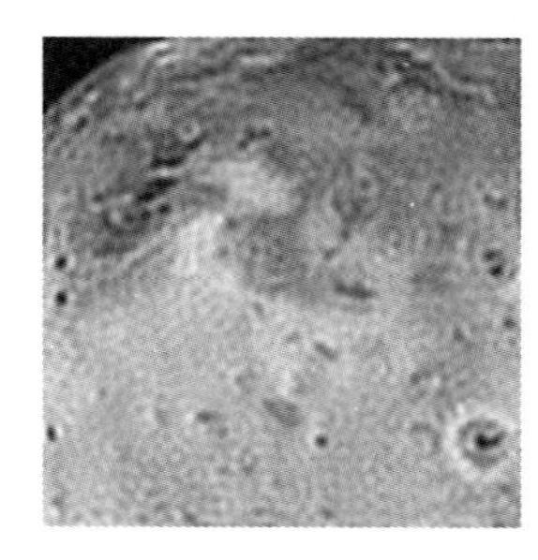

Figure 5.2. *TSVD solutions* $\mathbf{x}_k$ *using reflexive boundary conditions on a model problem created by the bordering technique (to ensure that the blurred image contains information from outside the edges). From left to right, we use truncation parameters* $k = 703$, $k = 2865$, *and* $k = 4638$.

Figure 5.2 shows reconstructions similar to those in Figure 5.1, except that we now use reflexive boundary conditions (leading to a matrix $\mathbf{A}$ with BTTB + BTHB + BHTB + BHHB structure; cf. Section 4.3). The bordering artifacts have disappeared.

5.3 SVD Analysis

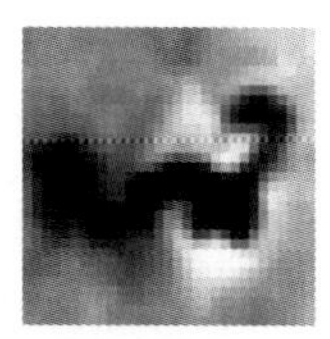

Figure 5.3. *The* 31×31 *test image* $\mathbf{X}$ *used in this section.*

In order to get a better understanding of the properties of the reconstructions computed by means of spectral filtering in the form of (5.3), this section consists of a case study for symmetric Gaussian blur (with $s_1 = s_2$ and $\rho = 0$). To simplify the discussion we keep the problem dimensions small and consider the test image $\mathbf{X}$ in Figure 5.3 with $m = n = 31$

(which is a subimage from the image `iogray.tif` showing a small detail), and we assume zero boundary conditions such that $\mathbf{A}_{\mathrm{BC}} = \mathbf{0}$ and $\mathbf{A} = \mathbf{A}_0$.

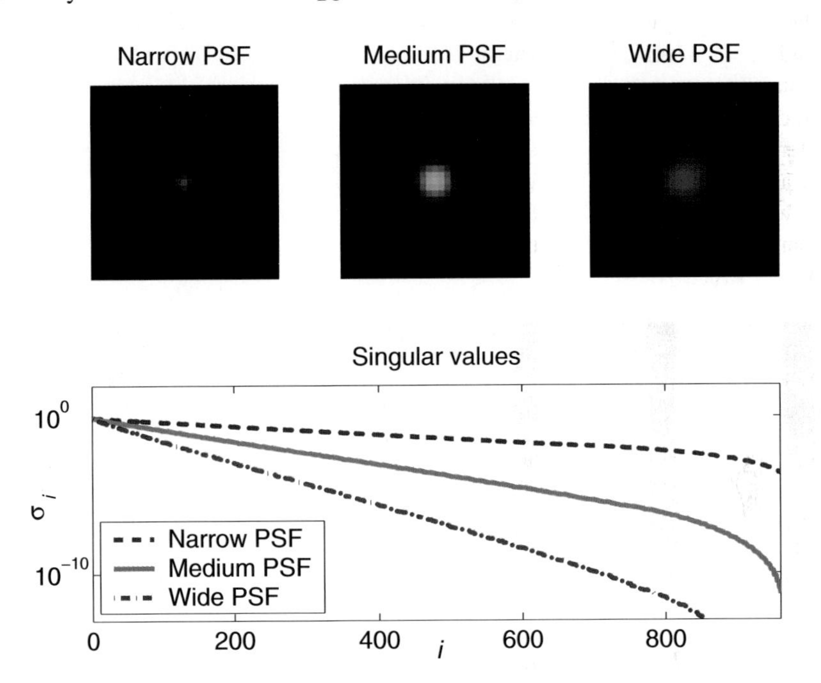

Figure 5.4. *Three Gaussian PSFs (with $s_1 = s_2 = 1$, 1.7, and 2.4, respectively) and the singular values σ_i of the corresponding blurring matrices $\mathbf{A} = \mathbf{A}_0$ (we assume zero boundary conditions). The decay of the singular values depends on the parameters s_1 and s_2 which define the width of the Gaussian function and, thus, the amount of blurring.*

Clearly, in spectral filtering we express the solution $\mathbf{x}_{\mathrm{filt}}$ as a sum of right singular vectors $\mathbf{v}_i$. The coefficients in this expansion are $\phi_i\, \mathbf{u}_i^T\mathbf{b} \,/\, \sigma_i$, where $\mathbf{u}_i^T\mathbf{b} \,/\, \sigma_i$ are the coefficients in the expansion of the naïve solution, and the filter factors ϕ_i are introduced to damp the undesired components. Hence we can obtain an understanding of the properties of $\mathbf{X}_{\mathrm{filt}}$ by studying the behavior of the quantities σ_i, $\mathbf{u}_i^T\mathbf{b}$, and $\mathbf{v}_i$.

Let us first look at the behavior of the singular values σ_i. We already mentioned in Chapter 1 that the singular values decay gradually. Figure 5.4 shows three 31×31 PSF arrays $\mathbf{P}$ together with the singular values of the corresponding 961×961 blurring matrices $\mathbf{A} = \mathbf{A}_0$. Notice that as the blurring gets worse—i.e., the PSF gets "wider"—the singular values decay faster. Even for narrow PSFs with a slow decay in singular values, the condition number $\mathrm{cond}(\mathbf{A}) = \sigma_1/\sigma_N$ becomes large for large images.

We note that at one extreme, when the PSF consists of a single nonzero pixel, the matrix $\mathbf{A}$ is the identity and all singular values are identical (and the condition number of the matrix is one). In the other extreme, when the PSF is so wide that all entries of $\mathbf{A}$ are equal, then all but one of the singular values are zero.

The behavior of the coefficients $\mathbf{u}_i^T\mathbf{b}$ is determined by both the blurring (via the singular vectors $\mathbf{u}_i$) and the data $\mathbf{b}$. Figure 5.5 shows plots of $|\mathbf{u}_i^T\mathbf{b}|$ for the same three

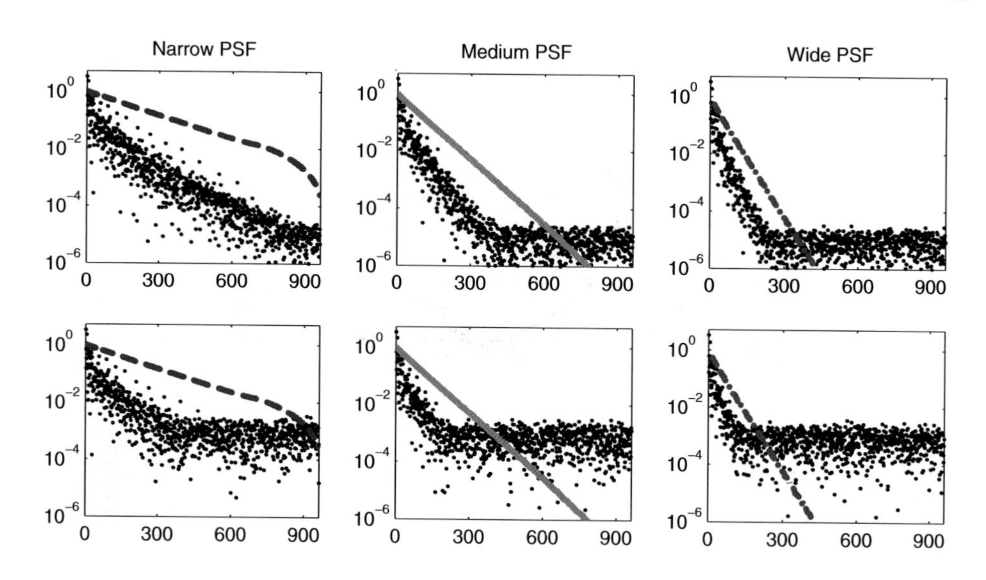

Figure 5.5. *Plots of singular values σ_i (colored lines) and coefficients $|\mathbf{u}_i^T\mathbf{b}|$ (black dots) for the three blurring matrices* $\mathbf{A}$ *defined by the PSFs in Figure* 5.4*, and two different levels of the noise* $\mathbf{E}$ *in the model* $\mathbf{B} = \mathbf{B}_{\text{exact}} + \mathbf{E}$*. Top row:* $\|\mathbf{E}\|_{\text{F}} = 3 \cdot 10^{-4}$*; bottom row:* $\|\mathbf{E}\|_{\text{F}} = 3 \cdot 10^{-2}$.

matrices $\mathbf{A}$ as in Figure 5.4, and for two different levels of the noise component $\mathbf{E}$ in the model $\mathbf{B} = \mathbf{B}_{\text{exact}} + \mathbf{E}$. We see from Figure 5.5 that, initially, the quantities $|\mathbf{u}_i^T\mathbf{b}|$ decay—at a rate that is slightly faster than that of the singular values—while later the coefficients level off at a noise plateau determined by the level of the noise in the image.

The insight we get from Figure 5.5 is that only some of the coefficients $\mathbf{u}_i^T\mathbf{b}$ carry clear information about the data, namely, those that are larger in absolute value than the noise plateau. Those coefficients lying at the noise plateau are dominated by the noise, hiding the true information. In other words, for the initial coefficients we have $\mathbf{u}_i^T\mathbf{b} \approx \mathbf{u}_i^T\mathbf{b}_{\text{exact}}$, while the remaining coefficients satisfy $\mathbf{u}_i^T\mathbf{b} \approx \mathbf{u}_i^T\mathbf{e}$. For any spectral filtering method, we therefore choose the filters ϕ_i so that the information in the initial coefficients dominates the filtered solution.

The index where the transition between the two types of behavior occurs depends on the noise level and the decay of the unperturbed coefficients. By a visual inspection of plots in Figure 5.5, we see that for $\|\mathbf{E}\|_{\text{F}} = 3 \cdot 10^{-4}$ the transition occurs very roughly around index 900 (narrow PSF), 400 (medium PSF), and 250 (wide PSF). Similarly, for the higher level of noise $\|\mathbf{E}\|_{\text{F}} = 3 \cdot 10^{-2}$ the transition occurs roughly around index 400 (narrow PSF), 200 (medium PSF), and 150 (wide PSF). These are precisely the values of k that should be used in the TSVD method.

If we choose the TSVD truncation parameter k too large, i.e., beyond the transition index for the coefficients $\mathbf{u}_i^T\mathbf{b}$, then we include, in addition to the desired SVD components, some components that are dominated by the noise. We say that such a reconstruction, in which we have included too many high-frequency SVD components, is undersmoothed.

If, on the other hand, we choose k too small, then we include too few SVD components, leading to a reconstruction that has a blurred appearance because it consists mainly of low-frequency information. A reconstruction with too few SVD components is said to be oversmoothed.

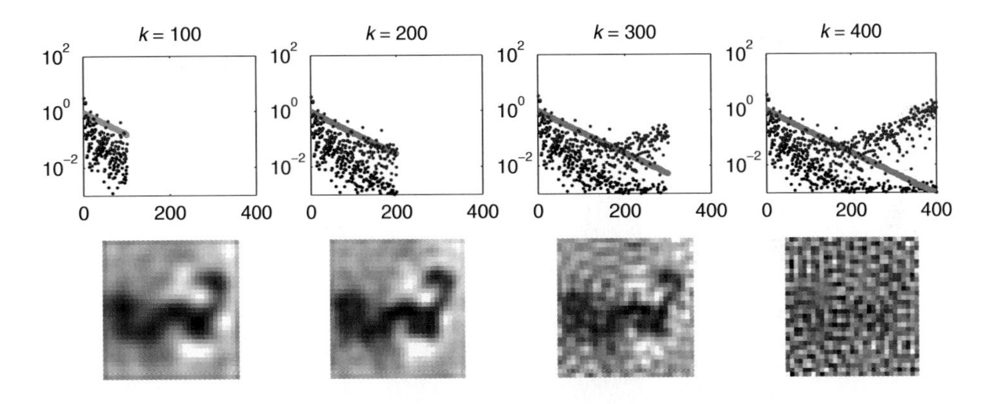

Figure 5.6. *Top row: singular values* σ_i *(green solid curve), right-hand side coefficients* $|\mathbf{u}_i^T \mathbf{b}|$ *(black dots), and TSVD solution coefficients* $|\mathbf{u}_i^T \mathbf{b}/\sigma_i|$ *(blue dots) for* $k = 100, 200, 300$, *and* 400. *Bottom row: the corresponding TSVD reconstructions. We use the medium PSF from Figure* 5.4, *and the higher level of noise* $\|E\|_F = 3 \cdot 10^{-2}$ *from Figure* 5.5.

To illustrate this, we use the problem in the middle of the bottom row in Figure 5.5 (the medium PSF and the noise level $3 \cdot 10^{-2}$), and we compute TSVD solutions $\mathbf{x}_k$ for $k = 100$, 200, 300, and 400. Figure 5.6 shows these solutions, along with plots of the singular values σ_i, the right-hand side coefficients $\mathbf{u}_i^T \mathbf{b}$, and the solution coefficients $\mathbf{u}_i^T \mathbf{b}/\sigma_i$. Clearly, $k = 200$ is the best of the four choices; for $k = 300$ some noise has entered the TSVD solution; and for $k = 400$ the TSVD solution is dominated by the inverted noise. On the other hand, $k = 100$ gives a blurred TSVD solution because too few SVD components are used.

VIP 15. The spectral components that are large in absolute value primarily contain pure data, while those with smaller absolute value are dominated by noise. The former components typically correspond to the larger singular values.

5.4 The SVD Basis for Image Reconstruction

Recall that we can always go back and forth between an image and its vector representation via the vec notation. Specifically, we can write

$$\mathbf{x}_{\text{filt}} = \text{vec}(\mathbf{X}_{\text{filt}}) \qquad \text{and} \qquad \mathbf{v}_i = \text{vec}\big(\mathbf{V}_i\big), \quad i = 1, \ldots, N.$$

That is, the image $\mathbf{X}_{\text{filt}}$ is the two-dimensional representation of the filtered solution $\mathbf{x}_{\text{filt}}$, and the basis images $\mathbf{V}_i$ are the two-dimensional representations of the singular vectors $\mathbf{v}_i$.

Using these quantities, we can thus write the filtered solution image as

$$\mathbf{X}_{\text{filt}} = \sum_{i=1}^{N} \phi_i \frac{\mathbf{u}_i^T \mathbf{b}}{\sigma_i} \mathbf{V}_i . \tag{5.5}$$

This expression is similar to (5.3), except that we write our reconstructed image $\mathbf{X}_{\text{filt}}$ as a weighted sum of the basis images $\mathbf{V}_i$. The expansion coefficients are the same as in (5.3), and we already studied them in Figures 5.5 and 5.6.

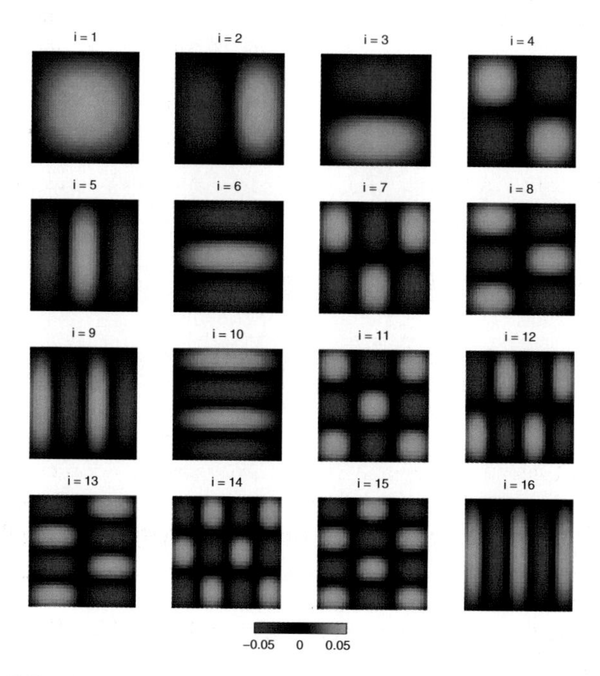

Figure 5.7. *Plots of the first* 16 *basis images* $\mathbf{V}_i$. *Green represents positive values in* $\mathbf{V}_i$ *while red represents negative values. These matrices satisfy* $\mathbf{v}_i = \text{vec}(\mathbf{V}_i)$, *where the singular vectors* $\mathbf{v}_i$ *are from the SVD of the matrix* $\mathbf{A}$ *for the middle PSF in Figure* 5.4.

Let us now look at the basis images $\mathbf{V}_i$ which constitute a basis for the filtered solution. Figure 5.7 shows $\mathbf{V}_i$ for $i = 1, \dots, 16$, arising from the SVD of the matrix $\mathbf{A}$ for the middle PSF in Figure 5.4. As the index i increases, the matrix $\mathbf{V}_i$ tends to have more high-frequency content, ranging from the very flat appearance of $\mathbf{V}_1$ to matrices with more oscillations in both the vertical and/or horizontal directions. This figure is another illustration of our claim in Chapter 1 that singular vectors corresponding to large singular values carry mainly low-frequency information, while singular vectors corresponding to the smaller singular values tend to represent more high-frequency information.

At this stage we recall the fact that if $\mathbf{A}$ is $N \times N$, then the singular values are uniquely determined. Unfortunately, the same is not true of the singular vectors, but this does not present any difficulties in our analysis of the spectral properties of the image deblurring problem. In particular, it can be shown that if the singular values, σ_i, are distinct, then the singular vectors are uniquely determined *up to signs*. In addition, the subspaces spanned by the singular vectors corresponding to distinct singular values are unique, and thus the singular vectors provide the necessary frequency information we desire. For symmetric Gaussian blur with $s_1 = s_2$, there are many identical pairs of singular values. For example, in Figure 5.7 only σ_1, σ_4, and σ_{11} are distinct, while the rest appear in pairs.

To further illustrate the role played by the SVD basis images $\mathbf{V}_i$, let us consider the same PSF but with either periodic or reflexive boundary conditions in the blurring model. That is, we now study the right singular vectors of the corrected matrix $\mathbf{A} = \mathbf{A}_0 + \mathbf{A}_{\mathrm{BC}}$, where $\mathbf{A}_{\mathrm{BC}}$ represents the correction corresponding to either periodic or reflexive boundary conditions. The first 16 matrices $\mathbf{V}_i$ for each boundary condition are shown in Figures 5.8 and 5.9; these basis images are quite different from those for zero boundary conditions (in Figure 5.7). This is due to the fact that each image $\mathbf{V}_i$ must satisfy the specified boundary conditions. Because of this, the reconstructed image will also be quite different.

VIP 16. The basis images $\mathbf{V}_i$ of $\mathbf{A}$ have increasingly more oscillations as the corresponding singular values σ_i decrease. In addition, each $\mathbf{V}_i$ satisfies the boundary conditions imposed on the problem.

5.5 The DFT and DCT Bases

The SVD gives us a general way to express the filtered solution in the forms (5.3) and (5.5). We also know from Chapter 4 that in many important cases we can compute the filtered solution efficiently via the FFT or the DCT, because these methods immediately provide the spectral factorization of the matrix $\mathbf{A}$. Hence, it is interesting to study the special basis components associated with these two methods.

Let us first consider the case where $\mathbf{A}$ is a BCCB matrix as described in Section 4.2. This matrix has the spectral decomposition $\mathbf{A} = \mathbf{F}^* \mathbf{\Lambda} \mathbf{F}$, in which $\mathbf{F}$ is the unitary two-dimensional DFT matrix. This matrix can be written as

$$\mathbf{F} = \mathbf{F}_{\mathrm{r}} \otimes \mathbf{F}_{\mathrm{c}},$$

where $\otimes$ is the Kronecker product defined in Section 4.4, and $\mathbf{F}_{\mathrm{c}}$ and $\mathbf{F}_{\mathrm{r}}$ are the unitary one-dimensional DFT matrices of dimensions $m \times m$ and $n \times n$, respectively. They are complex symmetric, $\mathbf{F}_{\mathrm{r}}^T = \mathbf{F}_{\mathrm{r}}$, $\mathbf{F}_{\mathrm{c}}^T = \mathbf{F}_{\mathrm{c}}$, and therefore

$$\mathbf{F}_{\mathrm{r}}^* = \mathrm{conj}(\mathbf{F}_{\mathrm{r}}), \qquad \mathbf{F}_{\mathrm{c}}^* = \mathrm{conj}(\mathbf{F}_{\mathrm{c}}),$$

in which $\mathrm{conj}(\cdot)$ denotes elementwise complex conjugation.

From the above expression for $\mathbf{F}$ it follows that

$$\mathbf{F}\,\mathbf{b} = \mathrm{vec}(\mathbf{F}_{\mathrm{c}}\,\mathbf{B}\,\mathbf{F}_{\mathrm{r}}^T),$$

in which the matrix $\mathbf{F}_{\mathrm{c}}\,\mathbf{B}\,\mathbf{F}_{\mathrm{r}}^T$ is identical (up to a scaling factor) to the two-dimensional FFT of $\mathbf{B}$. (This fact is used in VIP 10.) If we insert the relation for $\mathbf{F}\,\mathbf{b}$ into the expression for

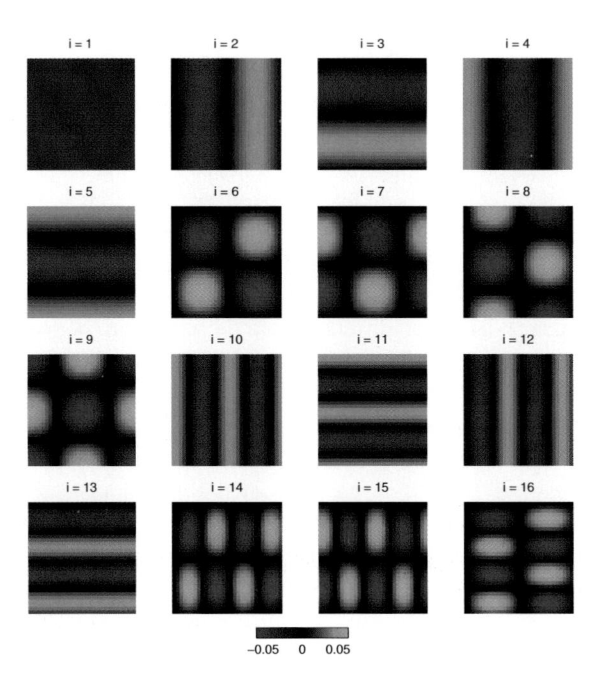

Figure 5.8. *The first* 16 *basis images* $\mathbf{V}_i$ *for the PSF in Figure* 5.7, *with* periodic *boundary conditions in the blurring model.*

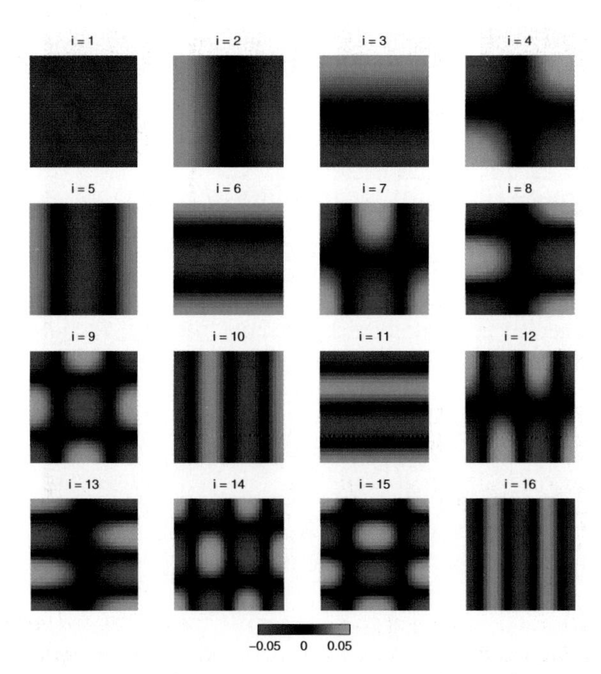

Figure 5.9. *The first* 16 *basis images* $\mathbf{V}_i$ *for the PSF in Figure* 5.7, *with* reflexive *boundary conditions in the blurring model.*

the naïve solution, we obtain

$$\mathbf{x}_{\text{naïve}} = \mathbf{A}^{-1}\mathbf{b} = \mathbf{F}^*\mathbf{\Lambda}^{-1}\mathbf{F}\,\mathbf{b} = (\mathbf{F}_\text{r} \otimes \mathbf{F}_\text{c})^*\mathbf{\Lambda}^{-1}\,\text{vec}(\mathbf{F}_\text{c}\,\mathbf{B}\,\mathbf{F}_\text{r}^T).$$

It is straightforward to show that if the $m \times n$ matrix $\mathbf{\Psi}$ satisfies

$$\text{vec}(\mathbf{\Psi}) = \mathbf{\Lambda}^{-1}\text{vec}(\mathbf{F}_\text{c}\,\mathbf{B}\,\mathbf{F}_\text{r}^T),$$

then the naïve reconstructed image is given by

$$\mathbf{X}_{\text{naïve}} = \mathbf{F}_\text{c}^*\,\mathbf{\Psi}\,(\mathbf{F}_\text{r}^*)^T = \text{conj}(\mathbf{F}_\text{c})\,\mathbf{\Psi}\,\text{conj}(\mathbf{F}_\text{r}^T) = \sum_{i=1}^{m}\sum_{j=1}^{n} \psi_{ij}\,\text{conj}(\mathbf{f}_{\text{c},i}\,\mathbf{f}_{\text{r},j}^T), \tag{5.6}$$

where ψ_{ij} are the elements of the matrix $\mathbf{\Psi}$. Moreover, $\mathbf{f}_{\text{c},i}$ is the ith column of $\mathbf{F}_\text{c}$, and $\mathbf{f}_{\text{r},j}$ is the jth column of $\mathbf{F}_\text{r}$. Equation (5.6) shows that in the BCCB case the basis images for the solution are the conjugate outer product $\text{conj}(\mathbf{f}_{\text{c},i}\,\mathbf{f}_{\text{r},j}^T)$ of all combinations of columns of the two DFT matrices. Figure 5.10 illustrates a few of these basis images.

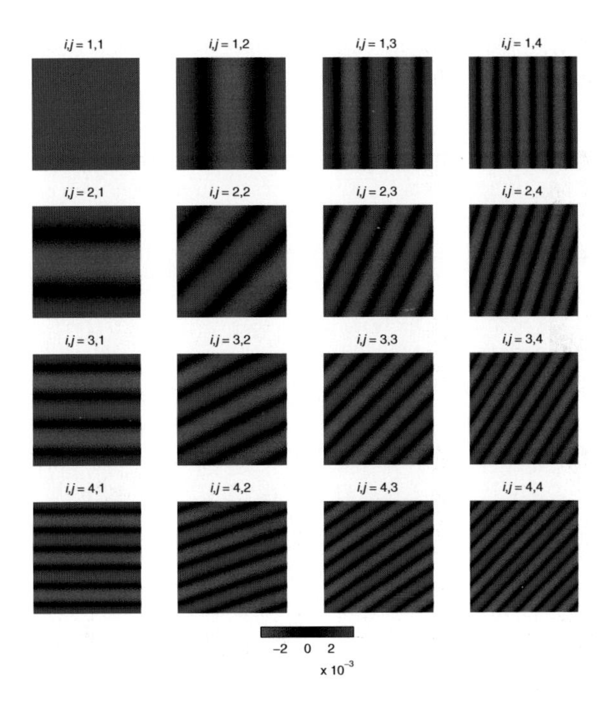

Figure 5.10. *Plots of the real parts of some of the DFT-based basis images* $\mathbf{f}_{\text{c},i}\,\mathbf{f}_{\text{r},j}^T$ *used when the blurring matrix* $\mathbf{A}$ *is a BCCB matrix; blue and red represent, respectively, positive and negative values. The imaginary parts of the basis images have the same appearance. We used* $m = n = 256$.

POINTER. The unitary DFT matrices $\mathbf{F}_c$ and $\mathbf{F}_r$ satisfy

$$\mathbf{F}_c = \text{conj}(\mathbf{F}_c)\,\mathbf{J}_c, \qquad \mathbf{F}_r = \text{conj}(\mathbf{F}_r)\,\mathbf{J}_r,$$

where $\mathbf{J}_c$ and $\mathbf{J}_r$ are permutation matrices of the form

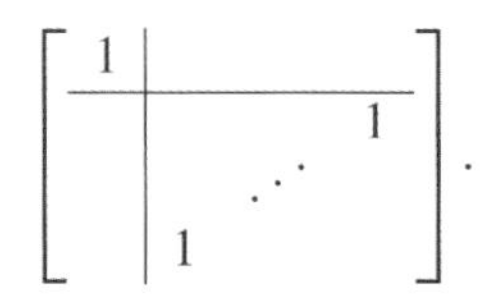

Since $\mathbf{B}$ is real, it follows that

$$\frac{1}{\sqrt{N}}\texttt{fft2}(\mathbf{B}) = \mathbf{F}_c\,\mathbf{B}\,\mathbf{F}_r^T = \mathbf{J}_c\,\text{conj}(\mathbf{F}_c\,\mathbf{B}\,\mathbf{F}_r^T)\,\mathbf{J}_r.$$

This symmetry is used in the `fft2` algorithm, when applied to real data, to halve the amount of computational work.

Next we consider the case from Section 4.3 where $\mathbf{A}$ is a sum of BTTB, BTHB, BHTB, and BHHB matrices. We can write $\mathbf{A} = \mathbf{C}^T\,\mathbf{\Lambda}\,\mathbf{C}$, where $\mathbf{C}$ is the two-dimensional DCT matrix. Similar to the BCCB case above, the matrix $\mathbf{C}$ can be be written as the Kronecker product

$$\mathbf{C} = \mathbf{C}_r \otimes \mathbf{C}_c,$$

where $\mathbf{C}_c$ and $\mathbf{C}_r$ are one-dimensional DCT matrices of dimensions $m \times m$ and $n \times n$, respectively. The rows of these matrices consist of sampled and scaled cosine functions. Inserting this into the expression for the naïve solution, and noticing that $\mathbf{C}\,\mathbf{b} = \text{vec}(\mathbf{C}_c\,\mathbf{B}\,\mathbf{C}_r^T)$, we obtain

$$\mathbf{x}_{\text{naïve}} = \mathbf{A}^{-1}\mathbf{b} = \mathbf{C}^T\mathbf{\Lambda}^{-1}\mathbf{C}\,\mathbf{b} = (\mathbf{C}_r \otimes \mathbf{C}_c)^T\,\mathbf{\Lambda}^{-1}\,\text{vec}(\mathbf{C}_c\,\mathbf{B}\,\mathbf{C}_r^T).$$

The matrix $\mathbf{C}_c\,\mathbf{B}\,\mathbf{C}_r^T$ is simply the two-dimensional DCT of the image $\mathbf{B}$.

In analogy with the DFT approach, we now introduce the $m \times n$ matrix Ψ such that

$$\text{vec}(\Psi) = \mathbf{\Lambda}^{-1}\text{vec}(\mathbf{C}_c\,\mathbf{B}\,\mathbf{C}_r^T),$$

and then we can write the naïve reconstructed image as

$$\mathbf{X}_{\text{naïve}} = \mathbf{C}_c^T\,\Psi\,\mathbf{C}_r = \sum_{i=1}^{m}\sum_{j=1}^{n} \psi_{ij}\,\mathbf{c}_{c,i}\,\mathbf{c}_{r,j}^T. \tag{5.7}$$

Here, $\mathbf{c}_{c,i}^T$ is the ith row of $\mathbf{C}_c$, and similarly $\mathbf{c}_{r,j}^T$ is the jth row of $\mathbf{C}_r$. Equation (5.7) shows that the basis images in this case are the outer product $\mathbf{c}_{c,i}\,\mathbf{c}_{r,j}^T$ of all combinations of rows of the two DCT matrices. Figure 5.11 shows a few of these basis images.

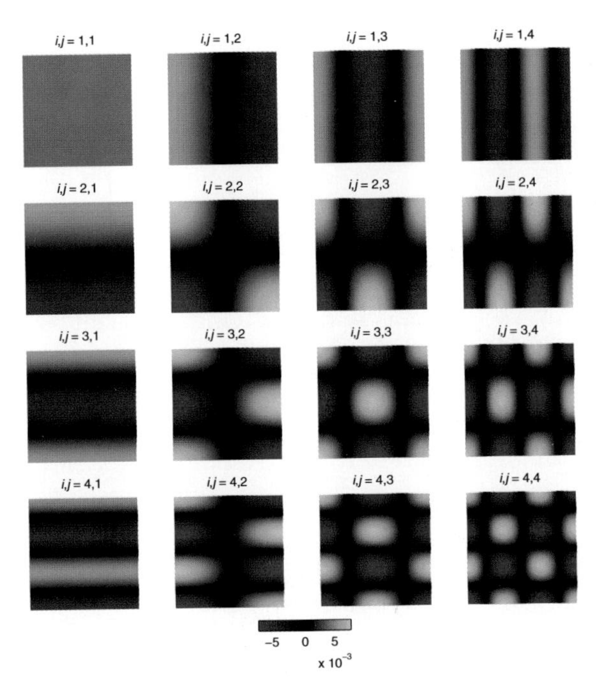

Figure 5.11. *Plots of some of the DCT-based basis images* $\mathbf{c}_{c,i}\,\mathbf{c}_{r,j}^T$ *used in the treatment of BTTB + BTHB + BHTB + BHHB matrices; cyan and red represent, respectively, positive and negative values. We used* $m = n = 256$.

5.6 The Discrete Picard Condition

Most images are dominated by low-frequency spectral components that are needed to describe the overall features of the image. High-frequency components are, of course, also needed to represent the details of the image, but these components are usually smaller in magnitude than the low-frequency components. This means that the expansion coefficients in a spectral basis will tend to decay in magnitude as the frequency increases.

We illustrate this fundamental property with an analysis of the `iogray.tif` image from Figure 5.1. For this image, we computed the three 512×512 arrays (or transforms)

$$\frac{1}{\sqrt{N}}\texttt{fft2}(\mathbf{X}) = \mathbf{F}_c\,\mathbf{X}\,\mathbf{F}_r^T, \qquad \texttt{dct2}(\mathbf{X}) = \mathbf{C}_c\,\mathbf{X}\,\mathbf{C}_r^T, \quad \text{and} \quad \mathbf{V}_c\,\mathbf{X}\,\mathbf{V}_r^T$$

which represent the spectral expansion coefficients in the DFT basis, the DCT basis, and the SVD basis for rotationally symmetric Gaussian blur (which is separable because $\rho = 0$). Excerpts of these arrays are shown in the top of Figure 5.12, where, for clarity, we have extracted the parts of the arrays that correspond to the lower frequencies. Note that we display the magnitudes of the coefficients.

In the DFT basis, the low-frequency information is represented by the coefficients in the four "corners" of the array, and we see that the largest coefficients (in magnitude) are

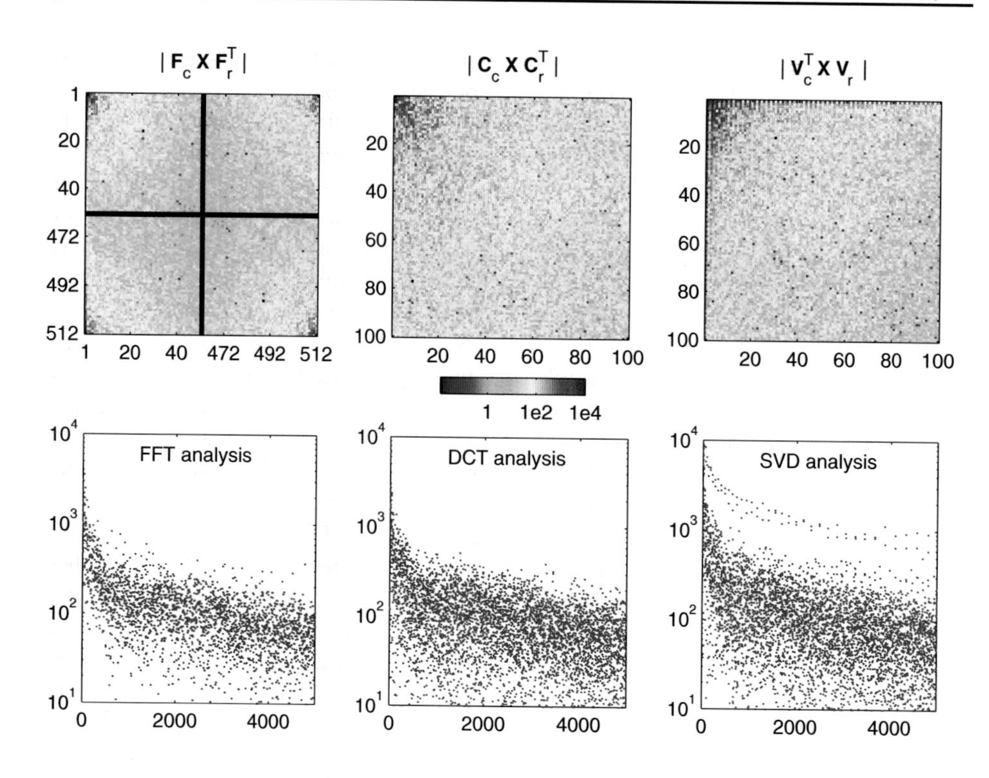

Figure 5.12. *Two representations of the magnitudes of the spectral components of the* `iogray.tif` *image in the DFT, DCT, and SVD bases for separable Gaussian blur. Top: the coefficients as they appear in the computed arrays. Bottom: the coefficients ordered according to decreasing eigenvalues or singular values.*

indeed located here. We also see that the magnitude of the coefficients tends to decay, on average, away from the four "corners" of the array. This verifies that the image is dominated by the low-frequency DFT basis images. Precisely the same is true in the DCT and the SVD bases, where the largest coefficients (in absolute value) are located in the upper left "corner" of the arrays—the image is also dominated by low-frequency DCT and SVD basis images.

The same information can also be represented or visualized in a different way, by plotting the spectral components (i.e., the elements of $\mathbf{F}_\mathrm{c}\mathbf{X}\mathbf{F}_\mathrm{r}^T$, $\mathbf{C}_\mathrm{c}\mathbf{X}\mathbf{C}_\mathrm{r}^T$, and $\mathbf{V}_\mathrm{c}^T\mathbf{X}\mathbf{V}_\mathrm{r}$) in the order dictated by a decreasing ordering of the corresponding eigenvalues or singular values of the blurring matrix $\mathbf{A}$. We have already seen such plots of the SVD coefficients in Figures 5.5 and 5.6. The eigenvalues and singular values are computed as explained in VIPs 10–12.

The bottom plots in Figure 5.12 show the first 5000 spectral components (in magnitude) according to this ordering, in which the smaller indices correspond to the lower frequencies. The overall tendency in all three plots is a decay of the magnitude of the spectral coefficients. The conclusion is therefore the same as before: the image $\mathbf{X}$ is dominated by the low-frequency spectral components corresponding to the largest eigenvalues or singular values, no matter which basis (DFT, DCT, or SVD) is used.

Now recall that the spectral coefficients of the exact blurred image $\mathbf{B}_{\text{exact}}$ (computed via $\mathbf{b}_{\text{exact}} = \mathbf{A}\,\mathbf{x}$) are obtained by multiplying the spectral coefficients for $\mathbf{X}$ by the eigenvalues or singular values of the blurring matrix $\mathbf{A}$. Due to the decay of the eigenvalues and singular values, it follows immediately that the spectral coefficients of the blurred image decay faster than those of the sharp image. This reflects the fact that high-frequency information is highly damped (or even lost) in the blurring process, and that blurred images are completely dominated by low-frequency components (giving the blurred appearance).

Concerning the reverse process of image *de*blurring, given a blurred and noisy image $\mathbf{B}$, the above analysis shows that we can only hope to compute an approximate reconstruction if the spectral components of $\mathbf{B}$ decay *faster* than the eigenvalues or singular values. This requirement to the data, or right-hand side $\mathbf{b}$, is known as the discrete Picard condition, and specifically for the SVD formulation this condition says that the right-hand side coefficients $|\mathbf{u}_i^T\mathbf{b}|$ must decay (on average) faster than the corresponding singular values.

Due to the presence of the noise in the recorded image $\mathbf{B} = \mathbf{B}_{\text{exact}} + \mathbf{E}$, we cannot expect all the coefficients $|\mathbf{u}_i^T\mathbf{b}|$ to decay, as is the case for the coefficients $|\mathbf{u}_i^T\mathbf{b}_{\text{exact}}|$. Rather the coefficients $|\mathbf{u}_i^T\mathbf{b}|$ will level off when they become dominated by the noise components. The index for which this transition occurs depends on the decay of the exact coefficients $|\mathbf{u}_i^T\mathbf{b}_{\text{exact}}|$ and the magnitude of the noise. When we compute an approximate reconstruction of the sharp image, we should include only the components that correspond to coefficients $|\mathbf{u}_i^T\mathbf{b}|$ that are above the noise level.

VIP 17. For most image deblurring problems, the SVD coefficients $|\mathbf{u}_i^T\mathbf{b}_{\text{exact}}|$ satisfy the discrete Picard condition. With the addition of noise, the coefficients $|\mathbf{u}_i^T\mathbf{b}|$ decay (on average) faster than the singular values σ_i, until they level off when the noise in the image starts to dominate the coefficients.

CHALLENGE 11. Smoothing (or blurring) acts as a low-pass filter that damps the higher frequencies in the image. High-pass filtering, which damps the lower frequencies, emphasizes the edges in the image. Use `conv2` to filter the `pumpkins.tif` image with a low-pass and a high-pass filter from Challenge 7, and inspect the three images. Then compute and display the two-dimensional DFT and/or the two-dimensional DCT of the image itself, as well as the filtered versions of the image (using `fft2` and/or `dct2`). You should see that the low-pass filter damps all the high frequencies, while the high-pass filter damps the low frequencies and magnifies higher frequencies.

CHALLENGE 12. Deblurring Walk-Through, Part I.

outoffocus1.tif

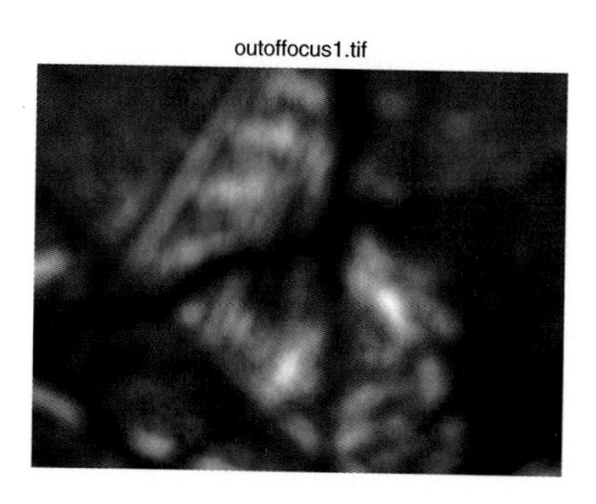

outoffocus2.tif

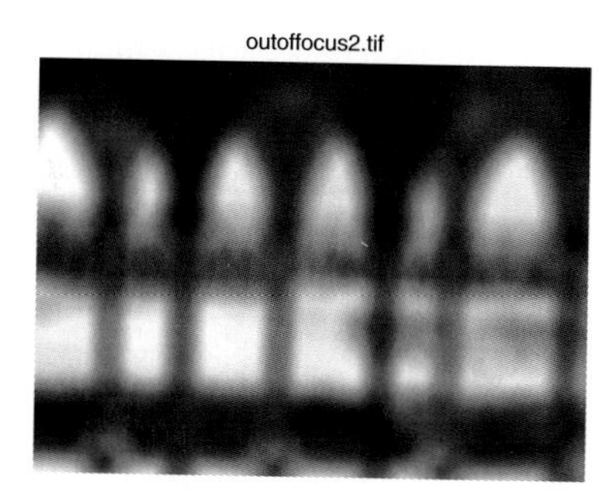

We provide two blurred images, shown above, both of which were degraded by out-of-focus blur. The PSF was generated by our `psfDefocus` function with `dim = 40` for `outoffocus1.tif` and `dim = 30` for `outoffocus2.tif`. The blurred images were rounded to integers in the range [0, 255], thus simulating quantization errors.
The best reconstructions for these images are obtained with reflexive boundary conditions. Your task here is to perform an analysis of the discrete Picard condition for one or both images. Specifically, you should plot the two-dimensional DCT coefficients for the blurred image (computed with `dct2` or `dcts2`) and the eigenvalues of the blurring matrix, computed as follows:

1. generate the small PSF array via a call to `psfDefocus`,
2. expand it to an $m \times n$ array via a call to `padPSF`, and
3. compute the eigenvalues as explained in VIP 11.

Use a semilogarithmic scale and plot the absolute values. You should see that the discrete Picard condition is satisfied for both images.
We return to the deblurring of these two images in Challenges 14, 15, and 18.

Chapter 6

Regularization by Spectral Filtering

You've got to go by or past or through boredom,
as through a filter, before the clear product emerges.
– F. Scott Fitzgerald

The previous chapter demonstrated that filtering is needed when noise is present. This chapter takes a closer look at filtering, which is also referred to as regularization because it can be interpreted as enforcing certain regularity conditions on the solution. The degree of regularization is governed by a regularization parameter that should be chosen carefully. We focus on two candidate regularization methods (TSVD and Tikhonov) and three candidate ways to compute the regularization parameter (the discrepancy principle, generalized cross validation, and the L-curve criterion).

6.1 Two Important Methods

The SVD analysis in the previous chapter motivates the use of spectral filtering methods because these methods give us control—via the filter factors—over the spectral contents of the deblurred images. Spectral filtering methods work by choosing the filter factors ϕ_i in the computed solution,

$$\mathbf{x}_{\text{filt}} = \sum_{i=1}^{N} \phi_i \frac{\mathbf{u}_i^T \mathbf{b}}{\sigma_i} \mathbf{v}_i , \tag{6.1}$$

POINTER. Many of the algorithms discussed in this chapter are implemented in the programs found in Appendices 1 and 2 and at the book's website. You can follow along with the codes as the algorithms are defined; see in particular

```
tik_dct    tik_fft    tik_sep    gcv_tik
tsvd_dct   tsvd_fft   tsvd_sep   gcv_tsvd
```

in order to obtain a solution with desirable properties. These methods operate on the data $\mathbf{b}$ in the coordinates $\mathbf{u}_i^T\mathbf{b}$ determined by the vectors $\mathbf{u}_i$ $(i = 1, \ldots, N)$ and express the solution $\mathbf{x}_{\text{filt}}$ in coordinates $\mathbf{v}_i^T\mathbf{x}$ determined by the vectors $\mathbf{v}_i$ $(i = 1, \ldots, N)$. This is the spectral coordinate system, since these vectors are the eigenvectors of $\mathbf{A}^T\mathbf{A}$ and $\mathbf{A}\mathbf{A}^T$, respectively.

We saw that solving the equation $\mathbf{A}\,\mathbf{x} = \mathbf{b}$ exactly did not produce a good solution when the data $\mathbf{b}$ was contaminated by noise. Instead, we filter the spectral solution via the filtered expansion in (5.3), so that the solution component in the direction $\mathbf{v}_i$ is scaled by the filter factor ϕ_i in order to reduce the effect of error in the component $\mathbf{u}_i^T\mathbf{b}$. In this section we discuss the two most important spectral filtering methods.

The TSVD Method. For this method, we define the filter factors to be one for large singular values, and zero for the rest. More precisely,

$$\phi_i \equiv \begin{cases} 1, & i = 1, \ldots, k, \\ 0, & i = k+1, \ldots, N. \end{cases} \tag{6.2}$$

The parameter k is called the truncation parameter and it determines the number of SVD components maintained in the regularized solution. Note that k always satisfies $1 \le k \le N$. This is the method used, for example, to compute the solution shown in Figure 5.6.

The Tikhonov Method. For this method we define the filter factors to be

$$\phi_i \equiv \frac{\sigma_i^2}{\sigma_i^2 + \alpha^2}, \qquad i = 1, \ldots, N, \tag{6.3}$$

where $\alpha > 0$ is called the regularization parameter. This choice of filter factors yields the solution vector $\mathbf{x}_\alpha$ for the minimization problem

$$\min_{\mathbf{x}} \left\{ \|\mathbf{b} - \mathbf{A}\,\mathbf{x}\|_2^2 + \alpha^2 \|\mathbf{x}\|_2^2 \right\}, \tag{6.4}$$

as we will discuss further in Section 7.2. This problem is motivated by the fact that we clearly want $\|\mathbf{b} - \mathbf{A}\,\mathbf{x}\|_2$ to be small, but if we make it zero by choosing $\mathbf{x} = \mathbf{A}^{-1}\mathbf{b}$, then

$$\|\mathbf{x}\|_2^2 = \sum_{i=1}^{N} \frac{(\mathbf{u}_i^T\mathbf{b})^2}{\sigma_i^2}.$$

This quantity becomes unrealistically large when the magnitude of the noise in some direction $\mathbf{u}_i$ greatly exceeds the magnitude of the singular value σ_i. Thus, we also want to keep $\|\mathbf{x}\|_2$ reasonably small, and our minimization problem in (6.4) ensures that both the norm of the residual $\mathbf{b} - \mathbf{A}\,\mathbf{x}_\alpha$ and the norm of the solution $\mathbf{x}_\alpha$ are somewhat small.

POINTER. We discuss the TSVD and the Tikhonov methods for choosing filter factors for deblurring. Filtering is a common task in signal processing, too. Rather than the SVD coordinate system $\mathbf{U}^T\mathbf{b}$, a Fourier coordinate system $\mathbf{Fb}$ is often used. Filters are chosen to diminish the effects of noise. A filter replaces a signal $\mathbf{b}$ by

$$\sum_{i=1}^{N} \phi_i (\mathbf{f}_i^T \mathbf{b}) \mathrm{conj}(\mathbf{f}_i),$$

where $\mathbf{f}_i^T$ is a row of the unitary Fourier transform matrix. For a low-pass filter, the filter factors for low-frequency components are close to one, while filter factors for high-frequency components are close to zero. The TSVD and Tikhonov methods are analogous to this, but the basis vectors are tailored to the blurring function. See [34] for more information about Fourier filtering.

We now consider the effect of the choice of the parameter α. Consider first a filter factor ϕ_i for which $\sigma_i \gg \alpha$ (which is the case for some of the first filter factors). Then, using the Taylor expansion $(1+\epsilon)^{-1} = 1 - \epsilon + \frac{1}{2}\epsilon^2 + O(\epsilon^3)$, we obtain

$$\phi_i = \frac{\sigma_i^2}{\sigma_i^2 + \alpha^2} = \frac{1}{1 + \alpha^2/\sigma_i^2} = 1 - \frac{\alpha^2}{\sigma_i^2} + \frac{1}{2}\frac{\alpha^4}{\sigma_i^4} + \cdots.$$

Next we consider a filter factor ϕ_i for which $\sigma_i \ll \alpha$ (which is the case for some of the last filter factors). Again using the Taylor expansion of $(1+\epsilon)^{-1}$, we obtain

$$\phi_i = \frac{\sigma_i^2}{\sigma_i^2 + \alpha^2} = \frac{\sigma_i^2}{\alpha^2}\frac{1}{1 + \sigma_i^2/\alpha^2} = \frac{\sigma_i^2}{\alpha^2}\left(1 - \frac{\sigma_i^2}{\alpha^2} + \frac{1}{2}\frac{\sigma_i^4}{\alpha^4} + \cdots\right).$$

Thus we can conclude that the Tikhonov filter factors satisfy

$$\phi_i = \begin{cases} 1 - \left(\dfrac{\alpha}{\sigma_i}\right)^2 + \mathcal{O}\left(\left(\dfrac{\alpha}{\sigma_i}\right)^4\right), & \sigma_i \gg \alpha, \\ \left(\dfrac{\sigma_i}{\alpha}\right)^2 + \mathcal{O}\left(\left(\dfrac{\sigma_i}{\alpha}\right)^4\right), & \sigma_i \ll \alpha. \end{cases}$$

This means that if we choose $\alpha \in [\sigma_N, \sigma_1]$, then $\phi_i \approx 1$ for small indices i, while $\phi_i \approx \sigma_i^2/\alpha^2$ for large i. For a given α, the "breakpoint" at which the filter factors change nature is at that index for which $\sigma_i \approx \alpha$. Figure 6.1 illustrates this point.

VIP 18. For TSVD regularization, we choose the truncation parameter k so that the residual $\|\mathbf{b} - \mathbf{Ax}\|_2$ is reasonably small, but the solution $\mathbf{x}$ does not include components corresponding to small singular values $\sigma_{k+1}, \ldots, \sigma_N$.
The parameter α in Tikhonov's method acts in the same way as the parameter k in the TSVD method: it controls which SVD components we want to damp or filter. We also see that there is no point in choosing α outside the interval $[\sigma_N, \sigma_1]$.

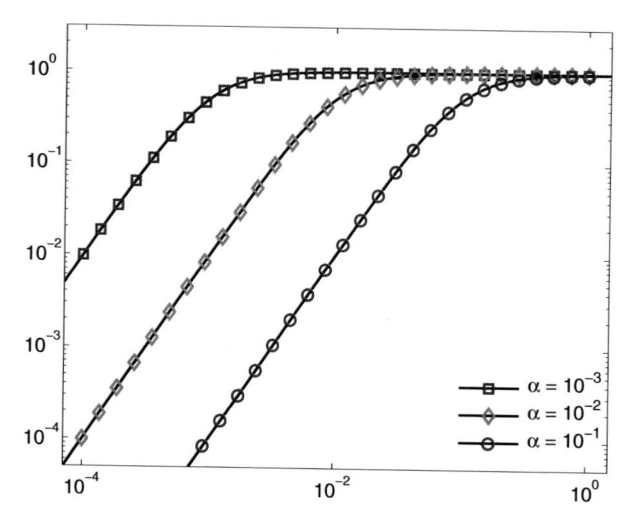

Figure 6.1. *The Tikhonov filter factors* $\phi_i = \sigma_i^2/(\sigma_i^2 + \alpha^2)$ *versus* σ_i *for three different values of the regularization parameter* α.

6.2 Implementation of Filtering Methods

If we assume that all of the singular values of **A** are nonzero, then the naïve solution can be written as

$$\mathbf{x} = \mathbf{A}^{-1}\mathbf{b} = \mathbf{V}\mathbf{\Sigma}^{-1}\mathbf{U}^T\mathbf{b}. \qquad (6.5)$$

Similarly, the spectral filter solution can be written as

$$\mathbf{x}_{\text{filt}} = \mathbf{V}\mathbf{\Phi}\mathbf{\Sigma}^{-1}\mathbf{U}^T\mathbf{b}, \qquad (6.6)$$

where $\mathbf{\Phi}$ is a diagonal matrix consisting of the filter factors ϕ_i for the particular method (e.g., 1's and 0's for TSVD and $\sigma_i^2/(\sigma_i^2 + \alpha^2)$ for Tikhonov). Relations analogous to (6.5) and (6.6) can be written in terms of the spectral decomposition, if it exists.

In Chapter 4 we described various kinds of structured matrices that arise in image deblurring problems, and we showed how to efficiently compute the SVD or spectral decomposition for such matrices. We also showed how to efficiently compute the naïve solution (6.5); recall VIPs 10, 11, and 12.

Since the expression (6.6) for $\mathbf{x}_{\text{filt}}$ is only a slight modification of (6.5), it is not difficult to efficiently implement filtering methods for the structured matrices of Chapter 4. Indeed, we can rewrite (6.6) as

$$\mathbf{x}_{\text{filt}} = \mathbf{V}\mathbf{\Sigma}_{\text{filt}}^{-1}\mathbf{U}^T\mathbf{b},$$

where $\mathbf{\Sigma}_{\text{filt}}^{-1} = \mathbf{\Phi}\mathbf{\Sigma}^{-1}$. Thus, if the filter factors are given, then it is simple to amend VIPs 10, 11, and 12 with implementations to compute $\mathbf{x}_{\text{filt}}$.

VIP 19. For many structured matrices, $\mathbf{x}_{\text{filt}}$ can be computed efficiently.

- Given:

  `P` = PSF
  `center = [row, col]` = center of PSF
  `B` = blurred image
  `BC` = string denoting boundary condition (e.g., `'zero'`)
  `Phi` = filter factors

- For periodic boundary conditions, use

```
S = fft2( circshift(P, 1 - center) );
Sfilt = Phi ./ S;
Xfilt = real( ifft2( fft2(B) .* Sfilt ) );
```

- For reflexive boundary conditions, with doubly symmetric PSF, use

```
e1 = zeros(size(P));, e1(1,1) = 1;
S = dct2( dctshift(P, center) ) ./ dct2(e1);
Sfilt = Phi ./ S;
Xfilt = idct2( dct2(B) .* Sfilt );
```

- For a separable PSF, use

```
[Ar, Ac] = kronDecomp(P, center, BC);
[Uc, Sc, Vc] = svd(Ac);
[Ur, Sr, Vr] = svd(Ar);
S = diag(Sc) * diag(Sr)';
Sfilt = Phi ./ S;
Xfilt = Vc * ( (Uc' * B * Ur) .* Sfilt ) * Vr';
```

We have said very little about how we choose the parameter k for TSVD or α for the Tikhonov method, except that the TSVD truncation parameter should satisfy $1 \le k \le N$, and the Tikhonov regularization parameter should satisfy $\sigma_N \le \alpha \le \sigma_1$. Later we discuss "automatic" methods for choosing these parameters, but for now we can try to choose them experimentally. For example, in the case of TSVD, we might specify a tolerance below which all singular (spectral) values are truncated. In this case the filter factors can be computed very easily in MATLAB as

```
Phi = (abs(S) >= tol);
```

By experimenting with various values of `tol`, and displaying the computed filtered solution, `Xfilt`, we can see the effects of regularization.

In the case of Tikhonov regularization, we could specify a value for α, and compute the filter factors from the singular (spectral) values as follows:

```
Phi = abs(S).^2 ./ (abs(S).^2 + alpha^2);
```

Note that the use of `abs` is necessary in the case when FFTs are used. Again, we can experiment with various values of `alpha` and display the filter solution to see the effects of regularization.

Figure 6.2. *The original and blurred pumpkin images (top), the Tikhonov reconstruction (bottom left), and the TSVD reconstruction (bottom right). The regularization parameters were chosen to give the visually most pleasing reconstruction.*

We illustrate in Figure 6.2 the results of the TSVD and Tikhonov methods on the blurred image of pumpkins from Figure 1.2. We used reflexive boundary conditions. For both methods, the regularization parameter was chosen to make the picture most pleasing to the eye. Both methods recover some detail from the blurred image, but TSVD tends to produce slightly more graininess for this image.

We close this section with a remark about computing the quantity

```
Sfilt = Phi ./ S.
```

If some of the singular (spectral) values in `S` are zero, then MATLAB will produce a "divide by zero" warning, and some values of `Sfilt` will be set to `Inf` or to `NaN`. We can avoid this unwanted situation by performing the computation only for nonzero values of `S`, and set all other `Sfilt` values to 0. To do this, we could, for example, use a logical array `idx = (S ~= 0)` that has values 1 for nonzero entries of `S` and 0 elsewhere. This array signals which divisions are to be performed, as illustrated in the following VIP.

VIP 20. Practical implementations of filtering methods should avoid possible divisions by zero. This can be done by using

```
idx = (S ~= 0);
Sfilt = zeros(size(Phi));
Sfilt(idx) = Phi(idx) ./ S(idx);
```

instead of the direct computation `Sfilt = Phi ./ S` used in VIP 19.

CHALLENGE 13. We return to the image in Challenge 2 and the blurring matrix in `challenge2.mat`. Use both the Tikhonov and the TSVD methods on this problem. Try various choices of the Tikhonov parameter α and the SVD truncation parameter k (using `tol` from p. 75) and determine which choices give the clearest solution image.

CHALLENGE 14. Deblurring Walk-Through, Part II. We return to the deblurring problem from Challenge 12 with out-of-focus blur and reflexive boundary conditions. For one or both images use the Tikhonov method implemented in `tik_dct` to compute and display reconstructions for selected values of the regularization parameter α. Note the following hints:

1. The original images have pixel values in the range $[0, 255]$.
2. Due to the reconstruction algorithm, some pixels in the reconstructions lie outside this range.
3. If you display a reconstruction `Xfilt` using the call `imshow(Xfilt,[ ])` or `imagesc(Xfilt)`, then MATLAB uses a color scale that maps `min(Xfilt(:))` to black and `max(Xfilt(:))` to white, leading to an image with different contrast than that in the original image.
4. To display the reconstruction with the same contrast as the original image, use `imshow(Xfilt,[0 255])` or `imagesc(Xfilt,[0 255])`.

We return to this deblurring problem in Challenges 15 and 18.

6.3 Regularization Errors and Perturbation Errors

In order to better understand the mechanisms and regularizing properties of the spectral filtering methods, we now take a closer look at the errors in the regularized solution $\mathbf{x}_{\text{filt}}$. We observed in (6.6) that we can write $\mathbf{x}_{\text{filt}}$ for both the TSVD solution and the Tikhonov solution in terms of the SVD. Equipped with this formulation, we can now easily separate the two different types of errors in a regularized solution, computed by means of spectral filtering. Specifically, we have

$$
\begin{aligned}
\mathbf{x}_{\text{filt}} &= \mathbf{V}\,\boldsymbol{\Phi}\,\boldsymbol{\Sigma}^{-1}\mathbf{U}^T\mathbf{b} \\
&= \mathbf{V}\,\boldsymbol{\Phi}\,\boldsymbol{\Sigma}^{-1}\mathbf{U}^T\mathbf{b}_{\text{exact}} + \mathbf{V}\,\boldsymbol{\Phi}\,\boldsymbol{\Sigma}^{-1}\mathbf{U}^T\mathbf{e} \\
&= \mathbf{V}\,\boldsymbol{\Phi}\,\boldsymbol{\Sigma}^{-1}\mathbf{U}^T\mathbf{A}\,\mathbf{x}_{\text{exact}} + \mathbf{V}\,\boldsymbol{\Phi}\,\boldsymbol{\Sigma}^{-1}\mathbf{U}^T\mathbf{e} \\
&= \mathbf{V}\,\boldsymbol{\Phi}\,\mathbf{V}^T\mathbf{x}_{\text{exact}} + \mathbf{V}\,\boldsymbol{\Phi}\,\boldsymbol{\Sigma}^{-1}\mathbf{U}^T\mathbf{e},
\end{aligned}
$$

and therefore the error in $\mathbf{x}_{\text{filt}}$ is given by

$$
\mathbf{x}_{\text{exact}} - \mathbf{x}_{\text{filt}} = (\mathbf{I}_N - \mathbf{V}\,\boldsymbol{\Phi}\,\mathbf{V}^T)\mathbf{x}_{\text{exact}} - \mathbf{V}\,\boldsymbol{\Phi}\,\boldsymbol{\Sigma}^{-1}\mathbf{U}^T\mathbf{e}. \tag{6.7}
$$

We see that the error consists of two contributions with different origins.

- Regularization Error. The first component $(\mathbf{I}_N - \mathbf{V}\,\boldsymbol{\Phi}\,\mathbf{V}^T)\mathbf{x}_{\text{exact}}$ is the regularization error which is caused by using a regularized inverse matrix $\mathbf{V}\,\boldsymbol{\Phi}\,\boldsymbol{\Sigma}^{-1}\mathbf{U}^T$ (instead of the inverse $\mathbf{A}^{-1} = \mathbf{V}\,\boldsymbol{\Sigma}^{-1}\mathbf{U}^T$) in order to obtain the filtering. The matrix $\mathbf{V}\,\boldsymbol{\Phi}\,\mathbf{V}^T$ describes the mapping between the exact solution and the filtered solution $\mathbf{x}_{\text{filt}}$. If $\boldsymbol{\Phi} = \mathbf{I}_N$, then the regularization error is zero, since $\mathbf{V}\mathbf{V}^T = \mathbf{I}_N$. The closer $\boldsymbol{\Phi}$ is to the identity, the smaller the regularization error.

- Perturbation Error. The second component $\mathbf{V}\,\boldsymbol{\Phi}\,\boldsymbol{\Sigma}^{-1}\mathbf{U}^T\mathbf{e}$ is the perturbation error which consists of the inverted and filtered noise. If $\boldsymbol{\Phi} = \mathbf{0}$, then the perturbation error is zero. When most of the filter factors are small (or zero), then the inverted noise is heavily damped (the perturbation error is small).

Changes in the regularization parameter change $\boldsymbol{\Phi}$ and the two kinds of errors. When too many filter factors ϕ_i are close to one, then the regularization error is small, but the perturbation error is large because inverted noise enters the solution—we say that the solution is undersmoothed. On the other hand, when too few filter factors are close to one, then the regularization error is large while the perturbation error is small—the solution is oversmoothed. A proper choice of the regularization parameter balances the two types of errors.

Figure 6.3 illustrates how the norms of the regularization error and the perturbation error vary with the regularization parameter. The problem is the same as in Figure 5.6, and we use TSVD as the regularization method. We see that the two types of errors are balanced for $k \approx 200$, which is consistent with the observation made from Figures 5.5 and 5.6.

VIP 21. Regularization by means of spectral filtering requires finding a suitable balance between the regularization error and the perturbation error by choosing the filter factors appropriately.

The reason we are able to compute regularized approximations to the exact solution, in spite of the large condition number, is that the spectral filtering is able to suppress much of the inverted noise while—at the same time—keeping the regularization error small. This is possible because the image deblurring problem satisfies the discrete Picard condition defined in the previous chapter; i.e., the exact right-hand side exhibits decaying expansion coefficients when expressed in the spectral basis.

As a consequence, the noise affects primarily the high-frequency components which are associated with the smaller singular values and which are damped by the spectral filtering method. What is left in the regularized solution is primarily the more low-frequency SVD components associated with the larger singular values, and these components are dominated by the contributions from the exact right-hand side.

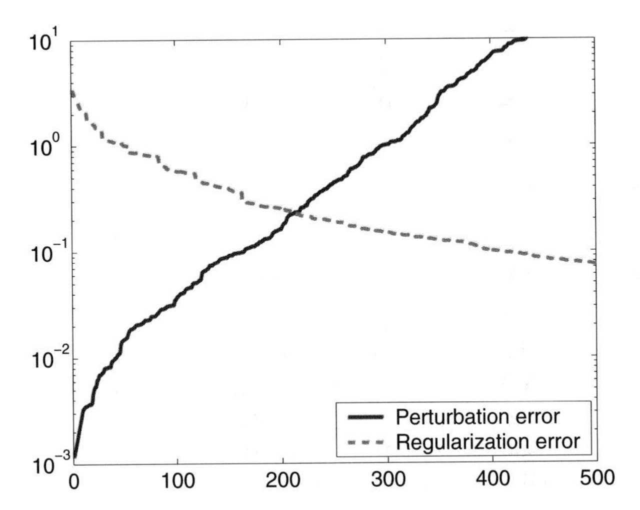

Figure 6.3. *The 2-norms of the regularization error* $(\mathbf{I}_N - \mathbf{V}\,\boldsymbol{\Phi}\,\mathbf{V}^T)\mathbf{x}_{\text{exact}}$ *and the perturbation error* $\mathbf{V}\,\boldsymbol{\Phi}\,\boldsymbol{\Sigma}^{-1}\mathbf{U}^T\mathbf{e}$ *versus the truncation parameter* k *for the TSVD method.*

To further understand this, let us consider the norm of the regularization error. Since $\mathbf{x}_{\text{exact}} = \mathbf{A}^{-1}\mathbf{b}_{\text{exact}} = \mathbf{V}\,\boldsymbol{\Sigma}^{-1}\mathbf{U}^T\mathbf{b}_{\text{exact}}$, and since $\|\mathbf{V}^T\mathbf{y}\|_2^2 = \|\mathbf{y}\|_2^2$, we obtain

$$\begin{aligned}\|(\mathbf{I}_N - \mathbf{V}\,\boldsymbol{\Phi}\,\mathbf{V}^T)\mathbf{x}_{\text{exact}}\|_2^2 &= \|(\mathbf{I}_N - \boldsymbol{\Phi})\,\mathbf{V}^T\mathbf{x}_{\text{exact}}\|_2^2 \\ &= \|(\mathbf{I}_N - \boldsymbol{\Phi})\,\boldsymbol{\Sigma}^{-1}\mathbf{U}^T\mathbf{b}_{\text{exact}}\|_2^2 \\ &= \sum_{i=1}^{N}\left((1-\phi_i)\,\frac{\mathbf{u}_i^T\mathbf{b}_{\text{exact}}}{\sigma_i}\right)^2 .\end{aligned}$$

Now recall that, due to the discrete Picard condition, the coefficients $|\mathbf{u}_i^T\mathbf{b}_{\text{exact}}/\sigma_i|$ decay (on average). Since the first filter factors ϕ_i (for $i = 1, 2, \ldots$) are close to one, the factors $(1-\phi_i)$ damp the contributions from the larger coefficients $\mathbf{u}_i^T\mathbf{b}_{\text{exact}}/\sigma_i$. Moreover, the small filter factors ϕ_i (for $i = N, N-1, \ldots$) correspond to factors $(1-\phi_i)$ close to one, which are multiplied by small coefficients $|\mathbf{u}_i^T\mathbf{b}_{\text{exact}}/\sigma_i|$. Hence we conclude that if the filters are suitably chosen, then the norm of the regularization error will not be large.

6.4 Parameter Choice Methods

Choosing the regularization parameter for an ill-posed problem is an art based on a combination of good heuristics and prior knowledge of the noise in the observations. We describe three important parameter choice methods: the discrepancy principle, generalized cross-validation, and the L-curve criterion.

For the discussion below, it is useful to have formulas for the norm of the spectral filtering solution

$$\|\mathbf{x}_{\text{filt}}\|_2^2 = \sum_{i=1}^{N}\left(\phi_i\,\frac{\mathbf{u}_i^T\mathbf{b}}{\sigma_i}\right)^2 \tag{6.8}$$

and the norm of the residual

$$\|\mathbf{b} - \mathbf{A}\,\mathbf{x}_{\text{filt}}\|_2^2 = \sum_{i=1}^{N} \left((1-\phi_i)\mathbf{u}_i^T\mathbf{b}\right)^2. \tag{6.9}$$

Note that for the TSVD method, the norm of the solution $\mathbf{x}_{\text{filt}} = \mathbf{x}_k$ is a monotonically nondecreasing function of k, while the residual norm is monotonically nonincreasing. Similarly, for the Tikhonov method, the norm of the solution $\mathbf{x}_{\text{filt}} = \mathbf{x}_\alpha$ is a monotonically nonincreasing function of α while the residual norm is monotonically nondecreasing.

The Discrepancy Principle [42]. This choice relies on having a good estimate of δ, the expected value of $\|\mathbf{e}\|_2$ (the error in the observations $\mathbf{b}$). If we have such an estimate, then the regularization parameter should be chosen so that the norm of the residual is approximately δ. Therefore, we choose a regularized solution $\mathbf{x}_{\text{filt}}$ so that

$$\|\mathbf{b} - \mathbf{A}\,\mathbf{x}_{\text{filt}}\|_2 = \tau\delta, \tag{6.10}$$

where $\tau > 1$ is some predetermined real number. (Common choices are $2 \le \tau \le 5$.) Note that as $\delta \to 0$, the filtered solution satisfies $\mathbf{x}_{\text{filt}} \to \mathbf{x}_{\text{exact}}$. Other methods based on knowledge of the variance are given, for example, in [5, 10, 19].

Since the residual norm is a monotonic function of our regularization parameter k or α, we can systematically try different values in (6.9) in order to find a parameter k or α to closely satisfy (6.10). Once the SVD, the filter factors, and the vector $\mathbf{U}^T\mathbf{b}$ have been computed, the cost is $2N$ multiplications and additions for each trial to compute the residual norm.

Generalized Cross Validation (GCV) [17]. This parameter choice method arises from the principle that if we omit a data value, then a good choice of the pixel values should be able to predict the missing data point well. The parameter is chosen to make the predictions as good as possible.

In contrast to the discrepancy principle, the parameter choice in GCV does not depend on a priori knowledge about the noise variance. Instead, GCV determines the parameter α that minimizes the GCV function, where α is the Tikhonov parameter or, abusing notation, $\alpha = 1/k$, where k is the TSVD cutoff. After a considerable amount of clever matrix manipulation, it can be shown that according to this principle, the best parameter for our spectral filtering methods minimizes the GCV functional

$$G(\alpha) = \frac{\|(\mathbf{I}_N - \mathbf{A}\,\mathbf{V}\,\mathbf{\Phi}\,\mathbf{\Sigma}^{-1}\mathbf{U}^T)\mathbf{b}\|_2^2}{(\text{trace}(\mathbf{I}_N - \mathbf{A}\,\mathbf{V}\,\mathbf{\Phi}\,\mathbf{\Sigma}^{-1}\mathbf{U}^T))^2}, \tag{6.11}$$

where, from the analysis in Section 6.3, $\mathbf{V}\,\mathbf{\Phi}\,\mathbf{\Sigma}^{-1}\mathbf{U}^T$ is the matrix that maps the right-hand side $\mathbf{b}$ onto the regularized solution $\mathbf{x}_{\text{filt}}$. GCV functions for the problem of Figure 6.3 are shown in Figure 6.4.

Although (6.11) is rather formidable, it is actually quite easy to evaluate. The numerator is just $\|(\mathbf{I}_N - \mathbf{A}\,\mathbf{V}\,\mathbf{\Phi}\,\mathbf{\Sigma}^{-1}\mathbf{U}^T)\mathbf{b}\|_2^2 = \|\mathbf{b} - \mathbf{A}\,\mathbf{x}_{\text{filt}}\|_2^2$, for which we already have a formula. We evaluate the denominator by noting that the trace of a matrix is the sum of its main diagonal elements, and the trace is invariant under orthogonal transformation, so

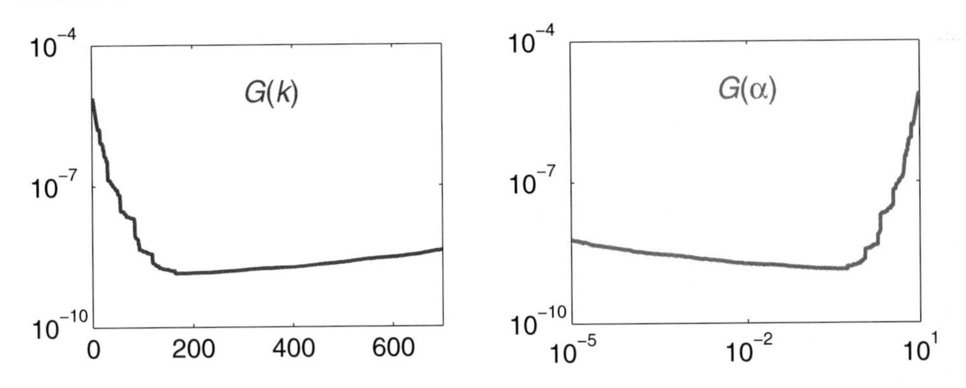

Figure 6.4. *The GCV functions* $G(k) = \|\mathbf{b} - \mathbf{A}\,\mathbf{x}_k\|_2^2/(N-k)^2$ *for TSVD (left) and* $G(\alpha)$ *given by* (6.11) *for Tikhonov regularization (right), applied to the same problem as in Figure* 6.3.

$$\begin{aligned}\text{trace}(\mathbf{I}_N - \mathbf{A}\,\mathbf{V}\,\mathbf{\Phi}\,\mathbf{\Sigma}^{-1}\mathbf{U}^T) &= \text{trace}(\mathbf{I}_N - \mathbf{U}\mathbf{\Sigma}\mathbf{V}^T\mathbf{V}\mathbf{\Phi}\mathbf{\Sigma}^{-1}\mathbf{U}^T)\\ &= \text{trace}(\mathbf{U}(\mathbf{I}_N - \mathbf{\Phi})\mathbf{U}^T)\\ &= \text{trace}(\mathbf{I}_N - \mathbf{\Phi})\\ &= N - \sum_{i=1}^{N}\phi_i\,,\end{aligned}$$

and this is easy to compute.

The L-Curve Criterion [20, 38]. The L-curve is a log-log plot of the norm of the regularized solution versus the corresponding residual norm for each of a set of regularization parameter values. For example, Figure 6.5 shows the L-curve for the TSVD method applied to the same problem as in Figure 6.3. This plot often is in the shape of the letter L, from which it draws its name. The log-log scale emphasizes the L shape.

Intuitively, the best regularization parameter should lie at the corner of the L, since for values higher than this, the residual increases rapidly and the norm of the solution decreases only slowly, while for values smaller than this, the norm of the solution increases rapidly without much decrease in the residual. Hence, we expect a solution near the corner to balance the regularization and perturbation errors.

In practice, only a few points on the L-curve need to be computed, and the corner is located by estimating the point of maximum curvature [24]. Computing a point on the L-curve costs only $3N$ multiplications and additions and N divisions using (6.8) and (6.9).

Which choice is best? Choosing an appropriate regularization parameter is very difficult. Every parameter choice method, including the three we discussed, has severe flaws: either they require more information than is usually available, or they fail to converge to the true solution as the error norm goes to zero. (For further discussion and references about parameter choice methods, see [10, 22, 36].)

- The discrepancy principle is convergent as the noise goes to zero, but it relies on information that is often unavailable or erroneous. Even with a correct estimate of the variance, the solutions tend to be oversmoothed [29, p. 96]. (See also the discussion in [20, Section 6.1].)

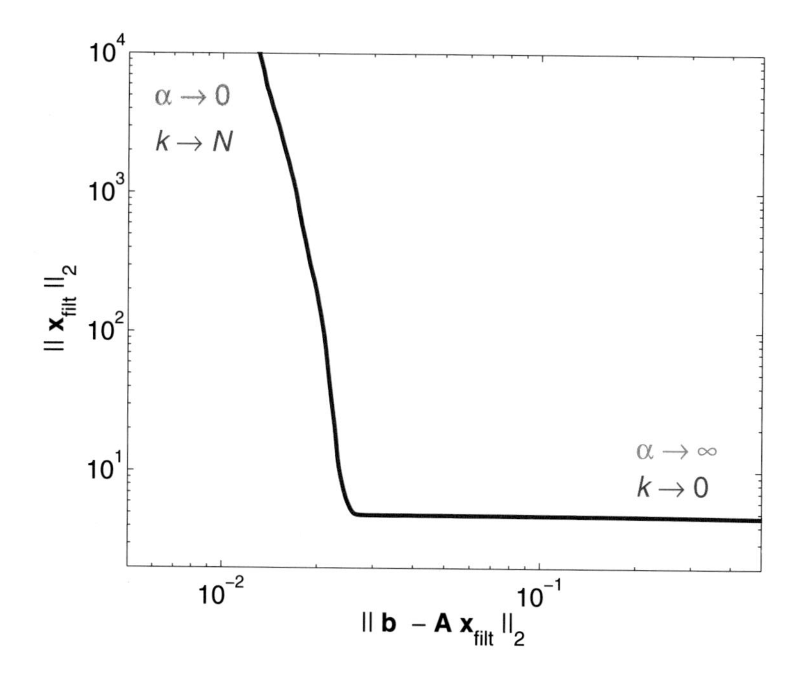

Figure 6.5. *The L-curve for TSVD with parameter k applied to the same problem as in Figure* 6.3, *i.e., a log-log plot of the solution norm* $\|\mathbf{x}_{\text{filt}}\|_2$ *versus the residual norm* $\|\mathbf{b} - \mathbf{A}\,\mathbf{x}_{\text{filt}}\|_2$. *The L-curve for Tikhonov regularization, with parameter* α, *would look similar.*

- For GCV, the solution estimates fail to converge to the true solution as the error norm goes to zero [12]. Another noted difficulty with GCV is that the graph for G can be very flat near the minimizer, so that numerical methods have difficulty in determining a good value of α [60].

- The L-curve criterion is usually more tractable numerically, but its limiting properties are far from ideal. The solution estimates fail to converge to the true solution as $N \to \infty$ [61] or as the error norm goes to zero [12].

VIP 22. No parameter choice method is perfect, and choosing among the discrepancy principle, GCV, the L-curve criterion, and other methods is dependent on what information is available about the problem.

6.5 Implementation of GCV

To use GCV to obtain an estimate of a regularization parameter, we must find the minimum of the function $G(\alpha)$ given in (6.11). In order to evaluate this function efficiently, some algebraic simplification is helpful. Specifically, in the case in which we are using the SVD, we obtain

$$G(\alpha) = \frac{\|(\mathbf{I}_N - \mathbf{A}\mathbf{V}\mathbf{\Phi}\mathbf{\Sigma}^{-1}\mathbf{U}^T)\mathbf{b}\|_2^2}{(\text{trace}(\mathbf{I}_N - \mathbf{A}\,\mathbf{V}\,\mathbf{\Phi}\,\mathbf{\Sigma}^{-1}\mathbf{U}^T))^2} = \frac{\|\mathbf{b} - \mathbf{A}\mathbf{x}_{\text{filt}}\|_2^2}{(\text{trace}(\mathbf{I}_N - \mathbf{\Phi}))^2}.$$

A similar simplification can be done for spectral decompositions, but for now we limit our discussion to the SVD. Observe that since the 2-norm is invariant under orthogonal transformation, $\|\mathbf{b} - \mathbf{A}\mathbf{x}_{\text{filt}}\|_2^2 = \|\mathbf{U}^T(\mathbf{b} - \mathbf{A}\mathbf{x}_{\text{filt}})\|_2^2$, so we can work in the coordinates of the SVD. Consider now specific regularization methods.

- GCV for TSVD. In the case of TSVD, the expression for $G(k)$ can be simplified further. In particular,

$$G(k) = \frac{1}{(N-k)^2} \sum_{i=k+1}^{N} (\mathbf{u}_i^T \mathbf{b})^2.$$

  Note that this is a discrete function. The truncation parameter is found by evaluating $G(k)$ for $k = 1, 2, \ldots, N-1$, and finding the index at which $G(k)$ attains its minimum.

- GCV for Tikhonov. In the case of Tikhonov filtering, the expression for $G(\alpha)$ becomes

$$G(\alpha) = \frac{\displaystyle\sum_{i=1}^{N} \left(\frac{\mathbf{u}_i^T \mathbf{b}}{\sigma_i^2 + \alpha^2} \right)^2}{\displaystyle\left(\sum_{i=1}^{N} \frac{1}{\sigma_i^2 + \alpha^2} \right)^2}.$$

  To find the minimum of this continuous function we can use MATLAB's built-in routine `fminbnd`. For example, suppose we implement the GCV function as

```
function G = GCV(alpha, bhat, s)
%  where bhat = U^T b.
phi_d = 1 ./ (s.^2 + alpha^2);
G = sum((bhat .* phi_d).^2)/(sum(phi_d)^2);
```

  Then the "optimal" α can be found using

```
alpha = fminbnd(@GCV,min(s),max(s),[ ],bhat,s);
```

where `s` = diag($\mathbf{\Sigma}$) and `bhat` = $\mathbf{U}^T\mathbf{b}$. If the spectral decomposition is used instead of the SVD, then some care must be taken in the implementation. Specifically, the values in `s` and `bhat` may be complex, and so absolute values must be included with the squaring operations. Further details on the implementations for the various structured matrices considered in Chapter 4 can be found in the program listings `gcv_tik` (Appendix 2) and `gcv_tsvd` (Appendix 1).

CHALLENGE 15. Deblurring Walk-Through, Part III. Once again we return to the problem from Challenges 12 and 14. For one or both images, compute the GCV function $G(\alpha)$ for the deblurring problem using reflexive boundary conditions and Tikhonov regularization. Then find the α that minimizes the GCV function and compute the corresponding reconstruction(s).

You should use the techniques discussed in this section to efficiently compute the GCV function. Keep in mind that for the algorithms based on the two-dimensional DCT, the SVD basis is replaced by the DCT (spectral) basis. Hence, you should replace $\mathbf{u}_i^T\mathbf{b}$ and σ_i^2 with $\mathbf{c}_i^T\mathbf{b}$ and λ_i^2, where $\mathbf{c}_i^T$ is the ith row of the two-dimensional DCT matrix $\mathbf{C}$ and λ_i is the ith eigenvalue of the blurring matrix; cf. Sections 4.3 and 5.4.

We return to this deblurring problem in Challenge 18.

6.6 Estimating Noise Levels

Further insight into choosing regularization parameters can be gained by using statistical information to estimate the noise level in the recorded image. Readers not familiar with statistics may wish to skip this section or consult an elementary statistics reference, such as [9, 52].

Let us consider the SVD analysis of the noise and the inverted noise. We first note that the coefficients $\mathbf{u}_i^T\mathbf{b}$ in the spectral expansion are the elements of the vector

$$\mathbf{U}^T\mathbf{b} = \mathbf{U}^T\mathbf{b}_{\text{exact}} + \mathbf{U}^T\mathbf{e}.$$

Let us now assume that the elements of the vector $\mathbf{e}$ are statistically independent, with zero mean and identical standard deviation (i.e., white noise). Then the expected value of $\mathbf{e}$ is the zero vector, while its covariance matrix is a scaled identity matrix,

$$\mathcal{E}(\mathbf{e}) = \mathbf{0}, \qquad \text{Cov}(\mathbf{e}) = \mathcal{E}(\mathbf{e}\,\mathbf{e}^T) = \eta^2\mathbf{I}_N,$$

where $\eta > 0$ is the standard deviation.

Then it follows from elementary statistics that the expected value of the vector $\mathbf{U}^T\mathbf{e}$ is also the zero vector, $\mathcal{E}(\mathbf{U}^T\mathbf{e}) = \mathbf{0}$, and that the covariance matrix for $\mathbf{U}^T\mathbf{e}$ is given by

$$\text{Cov}(\mathbf{U}^T\mathbf{e}) = \mathbf{U}^T\text{Cov}(\mathbf{e})\,\mathbf{U} = \eta^2\mathbf{U}^T\mathbf{U} = \eta^2\mathbf{I}_N.$$

Hence the coefficients $\mathbf{u}_i^T\mathbf{e}$ behave, statistically, like the elements of the noise vector $\mathbf{e}$. From elementary statistics we also obtain the following relation for the expected value of $(\mathbf{u}_i^T\mathbf{b})^2$:

$$\begin{aligned}\mathcal{E}\big((\mathbf{u}_i^T\mathbf{b})^2\big) &= \mathcal{E}\Big((\mathbf{u}_i^T\mathbf{b}_{\text{exact}} + \mathbf{u}_i^T\mathbf{e})^2\Big)\\ &= \mathcal{E}\Big((\mathbf{u}_i^T\mathbf{b}_{\text{exact}})^2 + 2\,\mathbf{u}_i^T\mathbf{b}_{\text{exact}}\,\mathbf{u}_i^T\mathbf{e} + (\mathbf{u}_i^T\mathbf{e})^2\Big)\\ &= (\mathbf{u}_i^T\mathbf{b}_{\text{exact}})^2 + \eta^2\end{aligned}$$

(because $\mathcal{E}(\mathbf{u}_i^T\mathbf{e}) = 0$), and the following first-order approximation holds:

$$\mathcal{E}\big(|\mathbf{u}_i^T\mathbf{b}|\big) \approx \sqrt{\mathcal{E}\big((\mathbf{u}_i^T\mathbf{b})^2\big)} = \sqrt{(\mathbf{u}_i^T\mathbf{b}_{\text{exact}})^2 + \eta^2}\,.$$

POINTER. The following relations hold for white Gaussian noise $\mathbf{e} \in \mathbb{R}^N$:

$$\mathcal{E}(e_i) = 0, \qquad \mathcal{E}(e_i^2) = \eta^2, \qquad \mathcal{E}(|e_i|) = \eta\,\sqrt{2/\pi} \approx \eta \cdot 0.8,$$

$$\mathcal{E}(\mathbf{e}) = \mathbf{0}, \qquad \mathcal{E}(\|\mathbf{e}\|_2^2) = \eta^2\, N, \qquad \mathcal{E}(\|\mathbf{e}\|_2) = \eta\,\frac{\sqrt{2}\,\Gamma((N+1)/2)}{\Gamma(N/2)}.$$

The factor $\sqrt{2}\,\Gamma((N+1)/2)/\,\Gamma(N/2)$ approaches $\sqrt{N}$ rapidly; for $N = 100$ the factor is 9.975. For more details see, e.g., Chapter 18 in [32].

We conclude that for any index i where $|\mathbf{u}_i^T \mathbf{b}_{\text{exact}}|$ is somewhat larger than η we have $\mathbf{u}_i^T \mathbf{b} \approx \mathbf{u}_i^T \mathbf{b}_{\text{exact}}$, while $\mathcal{E}(|\mathbf{u}_i^T \mathbf{b}|) \approx \eta$ when $|\mathbf{u}_i^T \mathbf{b}_{\text{exact}}|$ is smaller than η.

Returning to the plot in Figure 5.5, it is now evident that for small indices i the quantities $\mathbf{u}_i^T \mathbf{b}$ are indeed dominated by the component $\mathbf{u}_i^T \mathbf{b}_{\text{exact}}$ (which has an overall decreasing behavior), while for larger indices we have $\mathbf{u}_i^T \mathbf{b} \approx \mathbf{u}_i^T \mathbf{e} \approx \eta$, whose statistical behavior is identical to that of $\mathbf{e}$. We have thus explained the overall behavior of the plot.

This insight allows us to estimate the error in our data; $\eta \approx |\mathbf{u}_i^T \mathbf{b}|$ for large indices i. Once we have this estimate of the noise, we can, for example, use the discrepancy principle to guide our choice of TSVD or Tikhonov parameter.

CHALLENGE 16. Again consider the image in Challenge 2 and the blurring matrix in `challenge2.mat`. Use the discrepancy principle, GCV, and the L-curve criterion to choose regularization parameters for the Tikhonov method and the TSVD method on this problem. Compare the six parameters and the six solution images with the two that you computed in Challenge 13, in which you determined k and α to give the clearest solution image.

POINTER. Quantization noise for images represented as integers is uniformly distributed in the interval $[-a, a]$ with $a = 0.5$. The following relations hold for such quantization noise $\mathbf{e} \in \mathbb{R}^N$:

$$\mathcal{E}(e_i) = 0, \qquad \mathcal{E}(e_i^2) = \frac{a^2}{3}, \qquad \mathcal{E}(|e_i|) = \frac{a}{2},$$

$$\mathcal{E}(\mathbf{e}) = \mathbf{0}, \qquad \mathcal{E}(\|\mathbf{e}\|_2^2) = \frac{a^2}{3}\,N, \qquad \mathcal{E}(\|\mathbf{e}\|_2) = \gamma_N\, a\sqrt{\frac{N}{3}}.$$

The factor γ_N satisfies $\frac{1}{2}\sqrt{3} < \gamma_N < 1$, and it approaches 1 rapidly as N increases.

CHALLENGE 17. Use the approach described in Section 4.6 to create a blurred, noisy image with a Gaussian PSF. The functions

`tik_dct` `tik_fft` `tik_sep` `tsvd_dct` `tsvd_fft` `tsvd_sep`,

provided in Appendix 1 and Appendix 2 and available at the book's website, can be used to compute the Tikhonov and TSVD solutions with the various boundary conditions. Try each of these methods. In the case of `tik_sep` and `tsvd_sep`, experiment with different boundary conditions. In each case, use the default (GCV-chosen) regularization parameter, but also experiment with other values of the regularization parameter. For each method compare the quality of the reconstructed images, and the time required to compute them.

Why would you expect `tik_dct` and `tik_sep` with reflexive boundary conditions to compute similar results? Similarly, why would you expect `tik_fft` and `tik_sep` with periodic boundary conditions to compute similar results? Try a problem that has significant boundaries, as well as one that has zero boundaries. Which method would you recommend using for these particular problems?

CHALLENGE 18. Deblurring Walk-Through, Part IV. For the last time we return to the problem from Challenges 12, 14, and 15. For one or both images, compute the norm of the Tikhonov residual $\|\mathbf{b} - \mathbf{A}\,\mathbf{x}_{\text{filt}}\|_2$, determine the parameter α that satisfies the discrepancy principle (6.10) with $\tau = 2$, and compute the corresponding reconstruction(s). Compare with the GCV-based reconstruction. The residual norm can be computed via the same technique as used for the GCV function. To compute the estimate δ of the norm $\|\mathbf{e}\|_2$ of the errors in the data, we can assume that quantization errors dominate. Hence we can use $\delta = a\,\sqrt{mn/3}$.

Chapter 7

Color Images, Smoothing Norms, and Other Topics

Originality implies being bold enough to go beyond accepted norms.
– Anthony Storr

The purpose of this chapter is to introduce some more advanced topics related to spectral filtering methods. We start in Section 7.1 with a discussion of the treatment of color images, and we show how to deblur color images with the techniques introduced so far, with only minor changes. Next we show in Section 7.2 how to expand the use of the spectral filtering methods to more general regularization problems, where we incorporate a regularization term based on a smoothing norm, i.e., a norm that involves partial derivatives of the image. Section 7.3 discusses some computational issues when working with partial derivatives. We extend this minimization framework to other norms in Section 7.4, and to total variation techniques in Section 7.5. In Section 7.6 we discuss situations in which the blurring function is not known exactly. Finally, in Section 7.7, we discuss computational methods for use when either the size of the image or the properties of the PSF make it too expensive to use our spectral methods.

7.1 A Blurring Model for Color Images

As explained in Chapter 1, a digital color image is represented by a three-dimensional array `X` of size $m \times n \times 3$, in which the three two-dimensional arrays `X(:,:,1)`, `X(:,:,2)`, and `X(:,:,3)` represent the color information. The three arrays of the color image are often referred to as channels, and they often represent the intensities in the red, green, and blue scales. Figure 7.1 shows a color image and its three color channels.

To record a digital color image, one needs to filter the incoming light according to these colors before it is recorded. This can be done by first splitting the light beam in three directions, then filtering each of the three new beams by red, green, and blue filters, and finally recording these three color images on three different CCDs (charge-coupled devices). An alternative is to use a single CCD, in front of which is placed a filter mask with a red/green/blue pattern so that light in the different colors is recorded by different sensors on the same CCD. This filter, called a Bayer pattern, is illustrated in the left part of Figure 7.2.

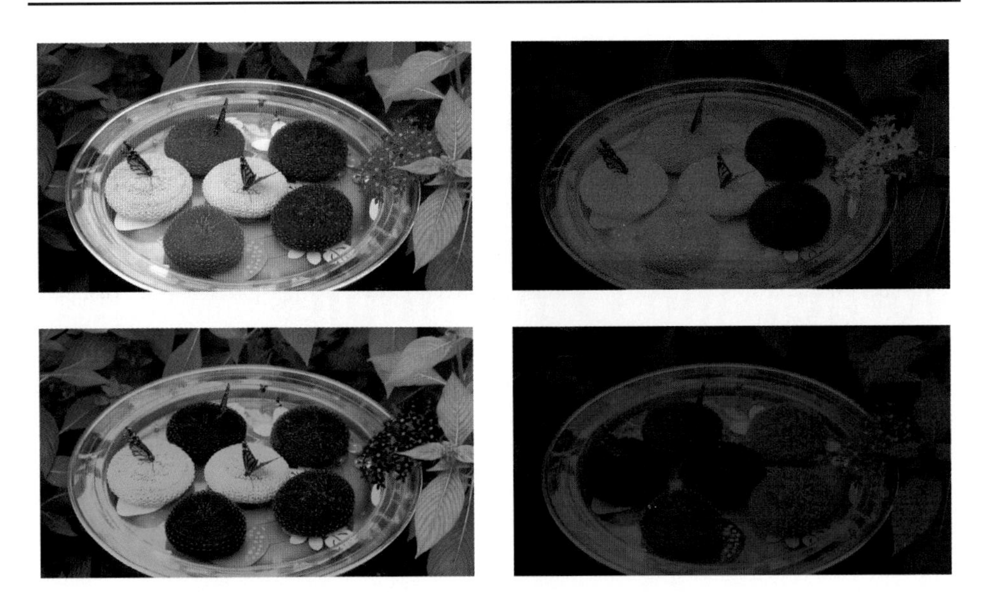

Figure 7.1. *A color image and its three color layers or channels.*

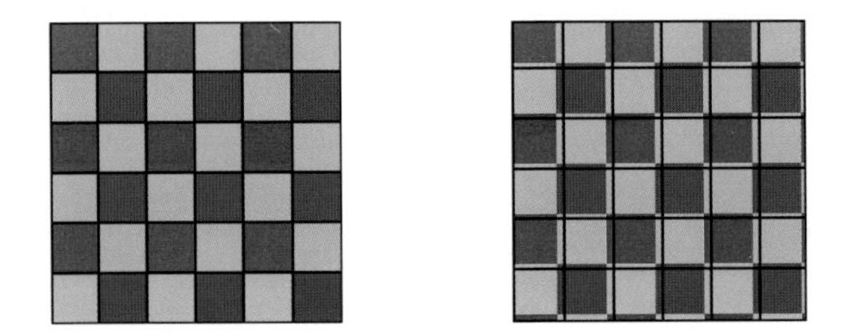

Figure 7.2. *Left: the Bayer pattern of sensors for recording color images on a single CCD; half of the sensors measure green intensities, while the others are evenly divided between red and blue. Right: the situation when the Bayer pattern is misaligned with respect to the CCD.*

In either technique for recording color images, there is a possibility that light from one color channel (red, green, or blue) ends up on a pixel assigned to another color. For example, in the single-CCD approach, this occurs if the Bayer pattern is misaligned with respect to the CCD, as shown in the right part of Figure 7.2. Due to the misalignment, a sensor meant to record green might be partially covered by the red mask instead of the green mask, so that it would be affected by red intensity as well as green. As a result, in addition to the within-channel blurring of each of the three color layers, there is also a cross-channel blurring among these layers.

Let $\mathbf{B}_{\mathrm{r}}$, $\mathbf{B}_{\mathrm{g}}$, and $\mathbf{B}_{\mathrm{b}}$ denote the grayscale images that constitute the three channels of the blurred color image $\mathbf{B}$. Assume that the color blurring (the cross-channel blurring) takes place after the optical blurring (the within-channel blurring) of the image. Then for each pixel, the red, green, and blue data vector that would be observed without the cross-channel blurring is multiplied by the matrix

$$\mathbf{A}_{\text{color}} = \begin{bmatrix} a_{\text{rr}} & a_{\text{rg}} & a_{\text{rb}} \\ a_{\text{gr}} & a_{\text{gg}} & a_{\text{gb}} \\ a_{\text{br}} & a_{\text{bg}} & a_{\text{bb}} \end{bmatrix}.$$

This matrix models the cross-channel blurring, and without loss of generality we can assume that each row sums to one.

Let us assume that the blurring is spatially invariant. This implies that the 3×3 cross-channel blurring matrix is the same for all pixels. In addition, we also assume that we have the same within-channel blurring (i.e., the same PSF) in all three channels; hence the optical blurring is modeled by

$$\mathbf{A}\,\text{vec}(\mathbf{X}_{\text{r}}), \qquad \mathbf{A}\,\text{vec}(\mathbf{X}_{\text{g}}), \qquad \text{and} \qquad \mathbf{A}\,\text{vec}(\mathbf{X}_{\text{b}}).$$

Then the model for the color blurring takes the block form

$$\begin{bmatrix} a_{\text{rr}}\mathbf{A} & a_{\text{rg}}\mathbf{A} & a_{\text{rb}}\mathbf{A} \\ a_{\text{gr}}\mathbf{A} & a_{\text{gg}}\mathbf{A} & a_{\text{gb}}\mathbf{A} \\ a_{\text{br}}\mathbf{A} & a_{\text{bg}}\mathbf{A} & a_{\text{bb}}\mathbf{A} \end{bmatrix} \begin{bmatrix} \text{vec}(\mathbf{X}_{\text{r}}) \\ \text{vec}(\mathbf{X}_{\text{g}}) \\ \text{vec}(\mathbf{X}_{\text{b}}) \end{bmatrix} = \begin{bmatrix} \text{vec}(\mathbf{B}_{\text{r}}) \\ \text{vec}(\mathbf{B}_{\text{g}}) \\ \text{vec}(\mathbf{B}_{\text{b}}) \end{bmatrix}$$

or, using Kronecker product notation,

$$\left(\mathbf{A}_{\text{color}} \otimes \mathbf{A}\right) \mathbf{x} = \mathbf{b}, \tag{7.1}$$

where, for color images, we define the "stacked" color images $\mathbf{b}$ and $\mathbf{x}$ by

$$\mathbf{b} = \begin{bmatrix} \text{vec}(\mathbf{B}_{\text{r}}) \\ \text{vec}(\mathbf{B}_{\text{g}}) \\ \text{vec}(\mathbf{B}_{\text{b}}) \end{bmatrix}, \qquad \mathbf{x} = \begin{bmatrix} \text{vec}(\mathbf{X}_{\text{r}}) \\ \text{vec}(\mathbf{X}_{\text{g}}) \\ \text{vec}(\mathbf{X}_{\text{b}}) \end{bmatrix}. \tag{7.2}$$

Figure 7.3. *Two types of blurred color images. Left: within-channel blurring only. Right: both within-channel blurring and cross-channel blurring with* $(a_{\text{rr}}, a_{\text{rg}}, a_{\text{rb}}) = (0.7, 0.2, 0.1)$, $(a_{\text{br}}, a_{\text{gg}}, a_{\text{gb}}) = (0.25, 0.5, 0.25)$, *and* $(a_{\text{br}}, a_{\text{bg}}, a_{\text{bb}}) = (0.15, 0.1, 0.75)$.

Figure 7.3 shows two blurred color images, one with within-channel blurring only and one with both within-channel and cross-channel blurring. Note how some of the color

POINTER. The technology associated with digital color imaging—including color fundamentals, color recording systems, and color image processing—is surveyed in [53].

information is absent from the latter image. In particular, if

$$\mathbf{A}_{\text{color}} = \begin{bmatrix} \frac{1}{3} & \frac{1}{3} & \frac{1}{3} \\ \frac{1}{3} & \frac{1}{3} & \frac{1}{3} \\ \frac{1}{3} & \frac{1}{3} & \frac{1}{3} \end{bmatrix},$$

then the resulting image is grayscale only, while, at the other extreme, no color blurring arises if $\mathbf{A}_{\text{color}} = \mathbf{I}_3$.

Deblurring of color images is done according to the color blurring model. If only within-channel blurring is present, then three independent deblurring problems are solved; otherwise the combined problem in (7.1) is solved.

VIP 23. Spatially invariant cross-channel (i.e., color) blurring is described by a 3×3 matrix $\mathbf{A}_{\text{color}}$ with nonnegative elements, and the full blurring model takes the form $\left(\mathbf{A}_{\text{color}} \otimes \mathbf{A}\right) \mathbf{x} = \mathbf{b}$.

7.2 Tikhonov Regularization Revisited

In Chapter 6 we mentioned that the Tikhonov solution is related to the minimization problem

$$\min_{\mathbf{x}} \left\{ \|\mathbf{b} - \mathbf{A}\,\mathbf{x}\|_2^2 + \alpha^2 \|\mathbf{x}\|_2^2 \right\},$$

whose solution has the form

$$\mathbf{x}_{\text{filt}} = \sum_{i=1}^{N} \frac{\sigma_i^2}{\sigma_i^2 + \alpha^2} \frac{\mathbf{u}_i^T \mathbf{b}}{\sigma_i} \mathbf{v}_i \,.$$

The above minimization problem turns out to be a special instance of a more general approach to image deblurring (in fact, a more general approach to the regularization of a wide class of inverse problems).

This general Tikhonov regularization or damped least squares method takes the form

$$\min_{\mathbf{x}} \left\{ \|\mathbf{b} - \mathbf{A}\,\mathbf{x}\|_2^2 + \alpha^2 \|\mathbf{D}\,\mathbf{x}\|_2^2 \right\}, \tag{7.3}$$

where $\mathbf{D}$ is a carefully chosen regularization matrix, often an approximation to a derivative operator. The interpretation of this minimization problem is that we seek a regularized solution that balances the size of two different terms:

- The first term $\|\mathbf{b} - \mathbf{A}\,\mathbf{x}\|_2^2$ is the square of the residual norm, and it measures the goodness-of-fit of the solution $\mathbf{x}$. If the residual is too large, then $\mathbf{A}\,\mathbf{x}$ does not fit the data $\mathbf{b}$ very well; on the other hand, if the residual is too small, then it is very likely that $\mathbf{x}$ is influenced too much by the noise in the data.

- The second term $\|\mathbf{D}\,\mathbf{x}\|_2^2$ is called the regularization term and involves a smoothing norm. This norm is the 2-norm of the solution when $\mathbf{D} = \mathbf{I}_N$. We choose $\mathbf{D}$ so that the regularization term is small when $\mathbf{x}$ matches our expectations of how a good quality solution should behave. We saw in Chapters 1 and 6 that it is the inverted noise that destroys the quality of the reconstruction, and therefore $\mathbf{D}$ should be chosen so that the regularization term is large when the reconstruction contains a large component of inverted noise.

The factor α^2 controls the balance between the minimization of these two quantities, and we saw in Chapter 6 that there are several methods available for choosing this factor. If α is too small, then we put too much emphasis on the first term, and $\mathbf{x}$ will be influenced too much by the noise in the data. On the other hand, if α is too large, then we put too much emphasis on the second term and thus obtain a very smooth solution with too few details.

VIP 24. Image deblurring can be formulated as an optimization problem of the form (7.3), where the goal is to control the size of both the goodness-of-fit term, $\|\mathbf{b} - \mathbf{A}\,\mathbf{x}\|_2^2$, and the regularization term, $\|\mathbf{D}\,\mathbf{x}\|_2^2$.

We now turn our attention to algorithmic considerations, while the choice of $\mathbf{D}$ is discussed in Section 7.3.

For the development of efficient algorithms for solving the Tikhonov problem, it is important to realize that (7.3) is a linear least squares problem. To see this, notice that for two vectors $\mathbf{y}$ and $\mathbf{z}$,

$$\|\mathbf{y}\|_2^2 + \|\mathbf{z}\|_2^2 = \mathbf{y}^T\mathbf{y} + \mathbf{z}^T\mathbf{z} = \begin{bmatrix}\mathbf{y}\\ \mathbf{z}\end{bmatrix}^T \begin{bmatrix}\mathbf{y}\\ \mathbf{z}\end{bmatrix} = \left\| \begin{bmatrix}\mathbf{y}\\ \mathbf{z}\end{bmatrix} \right\|_2^2 .$$

Hence, we can also write the Tikhonov problem in the form

$$\min_{\mathbf{x}} \left\| \begin{bmatrix}\mathbf{b}\\ \mathbf{0}\end{bmatrix} - \begin{bmatrix}\mathbf{A}\\ \alpha\mathbf{D}\end{bmatrix} \mathbf{x} \right\|_2^2 , \tag{7.4}$$

showing that it is, indeed, merely a linear least squares problem in $\mathbf{x}$. This is the best formulation for developing algorithms, since we can use numerical least squares algorithms to solve the Tikhonov problem efficiently. See, e.g., [4] and [22].

By setting the derivative of the minimization function in (7.4) to zero, it follows that the problem is mathematically equivalent to solving the normal equation

$$\left(\mathbf{A}^T\mathbf{A} + \alpha^2\mathbf{D}^T\mathbf{D}\right)\mathbf{x} = \mathbf{A}^T\mathbf{b}, \tag{7.5}$$

which is yet another formulation of the Tikhonov problem. From this equation we see that the Tikhonov solution $\mathbf{x}_{\alpha,\mathbf{D}}$ has a closed-form expression, namely,

$$\mathbf{x}_{\alpha,\mathbf{D}} = \left(\mathbf{A}^T\mathbf{A} + \alpha^2\mathbf{D}^T\mathbf{D}\right)^{-1}\mathbf{A}^T\mathbf{b}. \tag{7.6}$$

While (7.5) and (7.6) are convenient for theoretical studies, they are not recommended for numerical computations because of the loss of accuracy involved in working explicitly with the normal equations; cf. [28, Section 20.4] or [4, Section 2.2]. Instead we recommend using solution methods based directly on the least squares formulation (7.4), using a QR decomposition or an SVD.

VIP 25. There are three mathematically equivalent formulations (7.3)–(7.5) of the Tikhonov problem. All three are useful for various theoretical studies, but only the least squares formulation (7.4) is recommended for numerical computations.

7.3 Working with Partial Derivatives

In Chapter 6 we have already discussed the case where the squared 2-norm $\|\mathbf{x}\|_2^2$ provides an adequate indication of the "quality" of the solution $\mathbf{x}$. In terms of the general formulation in (7.3), this corresponds to choosing $\mathbf{D}$ to be the identity matrix. This is a good choice of regularization term whenever a large amount of inverted noise results in a large value of $\|\mathbf{x}\|_2$.

Other popular choices of the matrix $\mathbf{D}$ involve approximations to partial derivatives of the solution. The justification is that differentiation is known to magnify high-frequency components, and therefore we can often ensure the computation of a smooth solution by controlling the size of the partial derivatives.

Our main challenge in applying this method is to understand how we can compute approximations to the partial derivatives of a digital image. For ease of exposition we restrict ourselves to the case of periodic boundary conditions.

Let $\mathbf{z}$ be a vector of length m consisting of equidistant samples of a periodic function $z(t)$ with period 1, i.e.,

$$z_i = z(t_i) = z(t_{i+m}) \qquad \text{for} \quad i = 1, \ldots, m,$$

where $t_i = i\,h$ and $h = 1/m$ is the interspacing along the t-axis. Then the quantity $h^{-1}(z_{i+1} - z_i)$ is a finite-difference approximation to the first derivative $z'(t)$ at $t = t_i$. Similarly, the quantity $h^{-2}(z_{i+1} - 2z_i + z_{i-1})$ is a finite-difference approximation to the second derivative $z''(t)$ at $t = t_i$. If we define the two $m \times m$ banded circulant matrices $\mathbf{D}_{1,m}$ and $\mathbf{D}_{2,m}$ by

$$\mathbf{D}_{1,m} = \begin{bmatrix} -1 & 1 & & & 0 \\ 0 & -1 & 1 & & \\ & \ddots & \ddots & \ddots & \\ & & 0 & -1 & 1 \\ 1 & & & 0 & -1 \end{bmatrix} \tag{7.7}$$

and

$$\mathbf{D}_{2,m} = \begin{bmatrix} -2 & 1 & & & 1 \\ 1 & -2 & 1 & & \\ & \ddots & \ddots & \ddots & \\ & & 1 & -2 & 1 \\ 1 & & & 1 & -2 \end{bmatrix}, \tag{7.8}$$

then the vectors

$$h^{-1}\mathbf{D}_{1,m}\,\mathbf{z} \qquad \text{and} \qquad h^{-2}\mathbf{D}_{2,m}\,\mathbf{z}$$

contain the finite-difference approximations to $z'(t)$ and $z''(t)$ at the m points t_i.

Table 7.1. *Matrix expressions and corresponding computational stencils associated with discrete approximations of derivative operators. We let s denote the variable in the horizontal direction and let t denote the variable in the vertical direction. These derivatives can also be used in combination; for example, the Laplacian $x_{tt} + x_{ss}$ can be approximated by $h^{-2}(\mathbf{D}_{2,m}\mathbf{X} + \mathbf{X}\mathbf{D}_{2,n}^T)$.*

Matrix	$\mathbf{D}_{1,m}\mathbf{X}$	$\mathbf{D}_{2,m}\mathbf{X}$	$\mathbf{X}\mathbf{D}_{1,n}^T$	$\mathbf{X}\mathbf{D}_{2,n}^T$
Stencil	$\begin{bmatrix} 0 & 1 & 0 \\ 0 & -1 & 0 \\ 0 & 0 & 0 \end{bmatrix}$	$\begin{bmatrix} 0 & 1 & 0 \\ 0 & -2 & 0 \\ 0 & 1 & 0 \end{bmatrix}$	$\begin{bmatrix} 0 & 0 & 0 \\ 1 & -1 & 0 \\ 0 & 0 & 0 \end{bmatrix}$	$\begin{bmatrix} 0 & 0 & 0 \\ 1 & -2 & 1 \\ 0 & 0 & 0 \end{bmatrix}$
Approx.	$h\, x_t$	$h^2\, x_{tt}$	$h\, x_s$	$h^2\, x_{ss}$

In order to apply these ideas to images, we consider the pixels of the $m \times n$ digital image $\mathbf{X}$ as samples on an equidistant grid (with grid size $h \times h$) of a two-dimensional periodic function x, i.e., $x_{ij} = x(i\,h,\, j\,h)$. Now recall that we can always write $\mathbf{X}$ as a collection of rows or columns, i.e.,

$$\mathbf{X} = \begin{bmatrix} \mathbf{x}_1\,,\ \mathbf{x}_2\,,\ \ldots\,,\ \mathbf{x}_n \end{bmatrix} = \begin{bmatrix} \boldsymbol{\xi}_1^T \\ \vdots \\ \boldsymbol{\xi}_m^T \end{bmatrix},$$

where each column vector $\mathbf{x}_j$ consists of a column of pixels in the image, and each column vector $\boldsymbol{\xi}_i$ is chosen such that $\boldsymbol{\xi}_i^T$ is a row (or line) in the image. Then for $q = 1, 2$ the matrix

$$h^{-q}\mathbf{D}_{q,m}\mathbf{X} = h^{-q}\begin{bmatrix} \mathbf{D}_{q,m}\mathbf{x}_1\,,\ \mathbf{D}_{q,m}\mathbf{x}_2\,,\ \ldots\,,\ \mathbf{D}_{q,m}\mathbf{x}_n \end{bmatrix}$$

contains finite-difference approximations to the first or second partial derivative (x_t or x_{tt}) of the image in the vertical direction, evaluated at the pixel coordinates. Similarly, from the relation

$$h^{-q}\mathbf{X}\mathbf{D}_{q,n}^T = h^{-q}\begin{bmatrix} \boldsymbol{\xi}_1^T\mathbf{D}_{q,n}^T \\ \vdots \\ \boldsymbol{\xi}_m^T\mathbf{D}_{q,n}^T \end{bmatrix} = h^{-q}\begin{bmatrix} (\mathbf{D}_{q,n}\boldsymbol{\xi}_1)^T \\ \vdots \\ (\mathbf{D}_{q,n}\boldsymbol{\xi}_m)^T \end{bmatrix},$$

we see that the matrix $h^{-q}\mathbf{X}\mathbf{D}_{q,n}^T$ represents a finite-difference approximation to the first or second partial derivative (x_s or x_{ss}) of the image in the horizontal direction. Moreover, we see immediately that the matrix $h^{-2}(\mathbf{D}_{2,m}\mathbf{X}+\mathbf{X}\mathbf{D}_{2,n}^T)$ corresponds to the Laplacian $x_{tt} + x_{ss}$ of the image.

By writing out the elements of the matrices mentioned above we recognize the well-known 3×3 computational stencils listed in Table 7.1 for computing the discrete approximations to the five derivative operators. (A stencil is applied to the image through a convolution operation in the same way as a PSF array.) An example of the use of these stencils is shown

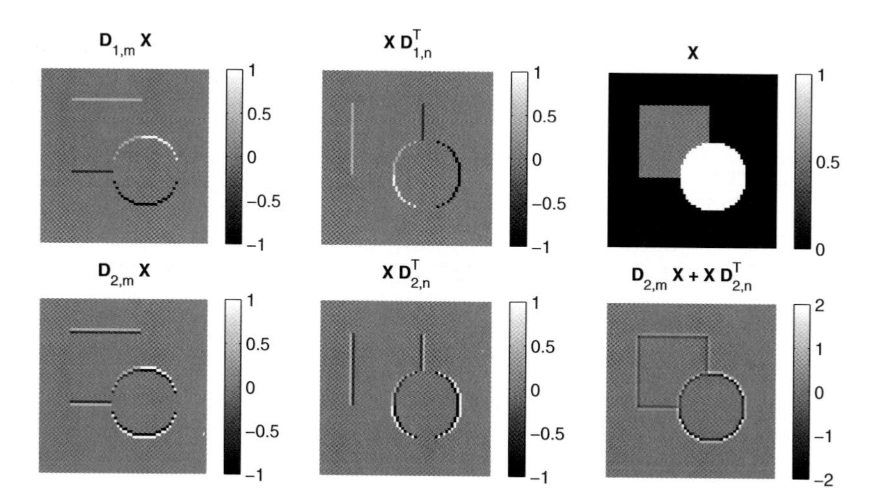

Figure 7.4. *The results of applying the partial derivative stencils from Table* 7.1 *to the test image from Challenge* 1. *Notice the different grayscales for the images.*

in Figure 7.4, where we see that the large elements in the matrices with derivatives ($\mathbf{D}_{q,m}\mathbf{X}$, etc.) correspond to the edges in the image $\mathbf{X}$.

Using the "vec" and Kronecker product notation from Section 4.4 and the relation $\|\mathbf{X}\|_\mathrm{F} = \|\mathrm{vec}(\mathbf{X})\|_2$, it follows that if $\mathbf{x} = \mathrm{vec}(\mathbf{X})$, then we can write the Frobenius norms of the discretized partial derivatives as h^{-q} times

$$\|\mathbf{D}_{q,m}\mathbf{X}\|_\mathrm{F} = \|\mathbf{D}_{q,m}\mathbf{X}\mathbf{I}_n\|_\mathrm{F} = \|(\mathbf{I}_n \otimes \mathbf{D}_{q,m})\mathbf{x}\|_2,$$

$$\|\mathbf{X}\mathbf{D}_{q,n}^T\|_\mathrm{F} = \|\mathbf{I}_m\mathbf{X}\mathbf{D}_{q,n}^T\|_\mathrm{F} = \|(\mathbf{D}_{q,n} \otimes \mathbf{I}_m)\mathbf{x}\|_2.$$

Similarly, the Frobenius norm of the discrete Laplacian of the image is h^{-2} times

$$\|\mathbf{D}_{q,m}\mathbf{X} + \mathbf{X}\mathbf{D}_{q,n}^T\|_\mathrm{F} = \|(\mathbf{I}_n \otimes \mathbf{D}_{q,m} + \mathbf{D}_{q,n} \otimes \mathbf{I}_m)\mathbf{x}\|_2.$$

We can also form a regularization term as follows:

$$\|(\mathbf{I}_n \otimes \mathbf{D}_{q,m})\,\mathbf{x}\|_2^2 + \|(\mathbf{D}_{q,n} \otimes \mathbf{I}_m)\,\mathbf{x}\|_2^2 = \left\| \begin{bmatrix} \mathbf{I}_n \otimes \mathbf{D}_{q,m} \\ \mathbf{D}_{q,n} \otimes \mathbf{I}_m \end{bmatrix} \mathbf{x} \right\|_2^2 .$$

There is no stencil formulation, similar to those in Table 7.1, for this norm.

The stage is now set for incorporating the partial derivatives, via the regularization terms, in the Tikhonov formulation of the image deblurring problems. Note that we can conveniently absorb the factor h^{-q} into the regularization parameter α. The different choices of $\mathbf{D}$ are summarized in Table 7.2.

VIP 26. Working with smoothing norms of the form $\|\mathbf{D}\mathbf{x}\|_2$ (instead of $\|\mathbf{x}\|_2$) adds only a minor overhead to the problem, namely, that of using one of the computational stencils shown in Table 7.1.

Table 7.2. *Matrices used in computing (scaled) derivative norms corresponding to different choices of* $\mathbf{D}$ *(periodic boundary conditions). The diagonal matrix* $\boldsymbol{\Delta}$ *is used in efficient implementation of the methods.*

Regularization term	Choice of $\mathbf{D}$	Choice of $\boldsymbol{\Delta}$
$\|x_s\|_2^2$ or $\|x_{ss}\|_2^2$	$\mathbf{I}_n \otimes \mathbf{D}_{q,m}$	$\mathbf{I}_n \otimes \|\boldsymbol{\Lambda}_{\mathbf{D}_{q,m}}\|^2$
$\|x_t\|_2^2$ or $\|x_{tt}\|_2^2$	$\mathbf{D}_{q,n} \otimes \mathbf{I}_m$	$\|\boldsymbol{\Lambda}_{\mathbf{D}_{q,n}}\|^2 \otimes \mathbf{I}_m$
$\|x_{ss} + x_{tt}\|_2^2$ (Laplacian)	$\mathbf{I}_n \otimes \mathbf{D}_{2,m} + \mathbf{D}_{2,n} \otimes \mathbf{I}_m$	$\mathbf{I}_n \otimes \boldsymbol{\Lambda}_{2,m}^2 + \boldsymbol{\Lambda}_{2,n}^2 \otimes \mathbf{I}_m$ $+ 2\, \boldsymbol{\Lambda}_{2,n} \otimes \boldsymbol{\Lambda}_{2,m}$
$\|x_s\|_2^2 + \|x_t\|_2^2$ or $\|x_{ss}\|_2^2 + \|x_{tt}\|_2^2$	$\begin{bmatrix} \mathbf{I}_n \otimes \mathbf{D}_{q,m} \\ \mathbf{D}_{q,n} \otimes \mathbf{I}_m \end{bmatrix}$	$\mathbf{I}_n \otimes \|\boldsymbol{\Lambda}_{\mathbf{D}_{q,m}}\|^2$ $+ \|\boldsymbol{\Lambda}_{\mathbf{D}_{q,n}}\|^2 \otimes \mathbf{I}_m$

POINTER. For periodic boundary conditions the DFT matrix $\mathbf{F} = \mathbf{F}_\mathrm{r} \otimes \mathbf{F}_\mathrm{c}$ diagonalizes the matrix $\mathbf{A}$, i.e.,

$$\mathbf{A} = \mathbf{F}^* \boldsymbol{\Lambda}_\mathbf{A}\, \mathbf{F},$$

where the diagonal matrix $\boldsymbol{\Lambda}_\mathbf{A}$ contains the eigenvalues of $\mathbf{A}$. The eigenvalues $\lambda_{q,i}$ of $\mathbf{D}_{q,m}$ are

$$\lambda_{q,k} = \begin{cases} \exp(2k\pi\hat{\imath}/m) - 1, & q = 1, \\ 2\cos(2k\pi/m) - 2, & q = 2 \end{cases} \qquad \text{for} \qquad k = 1, \dots, m,$$

in which $\hat{\imath} = \sqrt{-1}$ denotes the imaginary unit, and

$$\begin{aligned} \mathbf{I}_n \otimes \mathbf{D}_{q,m} &= \mathbf{F}^* (\mathbf{I}_n \otimes \boldsymbol{\Lambda}_{\mathbf{D}_{q,m}})\, \mathbf{F}, \\ \mathbf{D}_{q,n} \otimes \mathbf{I}_m &= \mathbf{F}^* (\boldsymbol{\Lambda}_{\mathbf{D}_{q,n}} \otimes \mathbf{I}_m)\, \mathbf{F}, \\ \mathbf{I}_n \otimes \mathbf{D}_{2,m} + \mathbf{D}_{2,n} \otimes \mathbf{I}_m &= \mathbf{F}^* (\mathbf{I}_n \otimes \boldsymbol{\Lambda}_{\mathbf{D}_{q,m}} + \boldsymbol{\Lambda}_{\mathbf{D}_{q,n}} \otimes \mathbf{I}_m) \mathbf{F}, \\ \begin{bmatrix} \mathbf{I}_n \otimes \mathbf{D}_{q,m} \\ \mathbf{D}_{q,n} \otimes \mathbf{I}_m \end{bmatrix} &= \begin{bmatrix} \mathbf{F}^* & 0 \\ 0 & \mathbf{F}^* \end{bmatrix} \begin{bmatrix} \mathbf{I}_n \otimes \boldsymbol{\Lambda}_{\mathbf{D}_{q,m}} \\ \boldsymbol{\Lambda}_{\mathbf{D}_{q,n}} \otimes \mathbf{I}_n \end{bmatrix} \mathbf{F}. \end{aligned}$$

For all four choices of $\mathbf{D}$ we can write the Tikhonov solution in the form

$$\mathbf{x}_{\alpha,\mathbf{D}} = \mathbf{F}^* \Big(|\boldsymbol{\Lambda}_\mathbf{A}|^2 \left(|\boldsymbol{\Lambda}_\mathbf{A}|^2 + \alpha^2\, \boldsymbol{\Delta} \right)^{-1} \Big)\, \boldsymbol{\Lambda}_\mathbf{A}^{-1}\, \mathbf{F}\, \mathbf{b}, \tag{7.9}$$

where the filtering matrix $|\boldsymbol{\Lambda}_\mathbf{A}|^2\, (|\boldsymbol{\Lambda}_\mathbf{A}|^2 + \alpha^2\, \boldsymbol{\Delta})^{-1}$ is diagonal. Here, the diagonal matrix $\boldsymbol{\Delta}$ takes one of the four forms shown in the last column of Table 7.2. The derivation of (7.9) can be found at the book's website.

POINTER. In the image processing literature, (7.9) is referred to as Wiener filtering [26]. The quantity $\alpha^2 \mathbf{\Delta}$ represents the noise-to-signal power, i.e., the power spectrum of the noise $\mathbf{E}$ divided by the power spectrum of the exact image $\mathbf{X}$. Wiener filtering is available in the IPT as the function `deconvwnr`.

POINTER. The IPT includes a function `deconvreg` that implements the algorithm described in this section for periodic boundary conditions. The user must specify information about the matrix $\mathbf{D}$ in the form of a stencil (cf. Table 7.1), and the default stencil is the Laplacian.

POINTER. Filtering using derivative operators and other smoothing norms is useful in, e.g., superresolution [13]. A survey of the use of derivative operators in image restoration can be found in [35].

7.4 Working with Other Smoothing Norms

The Tikhonov formulation in (7.3) is not the end of the story. There are many other possible choices of the regularization term. One way to generalize the regularization term is to replace the smoothing norm $\|\mathbf{D}\,\mathbf{x}\|_2^2$ with the norm $\|\mathbf{D}\,\mathbf{x}\|_p^p$, where $\|\cdot\|_p$ is the p-norm defined by

$$\|\mathbf{z}\|_p = \left(|z_1|^p + |z_2|^p + \cdots + |z_m|^p\right)^{1/p}.$$

Usually p satisfies $1 \le p \le 2$, because $p > 2$ leads to very smooth images that contain few details.

Table 7.3. *Illustration of the penalization of a large element by different choices of the norm $\|\cdot\|_p$. Here $\mathbf{z} = [\,3, 2, 1, 4\,]^T$ and $\hat{\mathbf{z}} = [\,3, 12, 1, 4\,]^T$.*

p	1	1.1	1.2	1.5	2
$\|\mathbf{z}\|_p^p$	10.0	11.1	12.3	17.0	30.0
$\|\hat{\mathbf{z}}\|_p^p$	20.0	24.3	29.7	57.8	170.0

While this may seem like only a minor change of the problem formulation, the impact on the reconstructed image can be dramatic. The reason is that a smoothing norm $\|\mathbf{D}\,\mathbf{x}\|_p^p$ with $p < 2$ penalizes the large-magnitude elements in the vector $\mathbf{D}\,\mathbf{x}$ less than the 2-norm does; and the smaller the p, the less the penalization. Table 7.3 illustrates this point with a small example. Hence, if we use a value of p close to one, then we can allow a larger fraction of not-so-small elements in $\mathbf{D}\,\mathbf{x}$ than when using $p = 2$.

Consider now the image deblurring problem in the form

$$\min_{\mathbf{x}} \left\{ \|\mathbf{b} - \mathbf{A}\,\mathbf{x}\|_2^2 + \alpha^2 \|\mathbf{D}\,\mathbf{x}\|_p^p \right\}, \tag{7.10}$$

where $\mathbf{D}$ is one of the matrices discussed above. If we use $p = 2$, then the smoothing term $\|\mathbf{D}\,\mathbf{x}\|_2^2$ enforces a dramatic penalization of large elements in the vector $\mathbf{D}\,\mathbf{x}$, which is equivalent to requiring that the partial derivatives must be small everywhere in the image. Thus, this approach favors reconstructions that are very smooth.

However, if we use a smoothing norm $\|\mathbf{D}\,\mathbf{x}\|_p^p$ with p close to one, then we implicitly allow the partial derivatives to be larger in certain limited regions of the image. This allows the large values of partial derivatives typically found near edges and discontinuities in an image. As a result of using p close to one, we can therefore obtain reconstructed images with better defined edges, because the large partial derivatives associated with the edges have less contribution to the function to be minimized.

We illustrate this with a color-image example with no cross-channel blurring. Each of the three color layers in the image is reconstructed by means of the smoothing norm $\|\mathbf{D}\,\mathbf{x}\|_p^p$ with the matrix

$$\mathbf{D} = \begin{bmatrix} \mathbf{I}_n \otimes \mathbf{D}_{1,m} \\ \mathbf{D}_{1,n} \otimes \mathbf{I}_m \end{bmatrix}, \tag{7.11}$$

which corresponds to penalizing the norm of the first-order partial derivatives in the horizontal and vertical directions in the image. Figure 7.5 shows two reconstructions using $p = 2$ and $p = 1.1$, respectively. Both reconstructions have sharp edges, but the second is free of the "freckles" that are clearly visible in the 2-norm reconstruction.

The Tikhonov problem based on a smoothing norm $\|\mathbf{D}\,\mathbf{x}\|_p^p$ with $p \neq 2$ is much more expensive computationally. The linear least squares algorithms are no longer directly applicable. However, it is still sometimes possible to make use of the fast methods described in this book. For example, the 1-norm and the ∞-norm formulations lead to linear programming problems, and these problems can be solved by solving a sequence of least squares problems [48]. A similar algorithm applies to various iteratively reweighted least squares problems [40, 45].

7.5 Total Variation Deblurring

The case $p = 1$ (the smallest value of p for which $\|\cdot\|_p$ is a norm) has received attention in connection with total variation denoising and deblurring of images, when it is used together with first derivatives. Here we give a brief introduction to this subject, expressed in our matrix terminology.

The gradient magnitude $\gamma(s,t)$ of the two-dimensional function $x = x(s,t)$ is a new two-dimensional function, defined as

$$\gamma(s,t) = \|\nabla x\|_2 = \left\| \begin{bmatrix} x_s \\ x_t \end{bmatrix} \right\|_2,$$

where again we use the notation x_s and x_t to denote, respectively, partial derivatives of x

Figure 7.5. *Comparison of image deblurring using the* `butterflies.tif` *color image from Figure 7.1. Both reconstructions were computed using the smoothing norm* $\|\mathbf{D}\,\mathbf{x}\|_p^p$ *with the matrix in (7.11). The top image, which was computed with* $p = 2$*, has sharp edges, but it also includes a large number of "freckles" with high spatial frequencies. The bottom image, which uses p-norm smoothing with* $p = 1.1$*, has no freckles, yet the sharp edges are reconstructed well.*

with respect to s and with respect to t. The total variation functional is then defined as

$$J_{\text{TV}}(x) = \|\gamma\|_1 = \int_0^1 \int_0^1 |\gamma(s,t)|\, ds\, dt.$$

From the example images in Figure 7.4 we see that if the function $x(s,t)$ represents an image, then the total variation functional $J_{\text{TV}}(x)$ is a measure of the accumulated size of the edges in the image.

POINTER. More details about total variation denoising and deblurring can be found in the books by Chan and Shen [7] and Vogel [62].

In the discrete formulation, the total variation of the image **X** is given by

$$J_{\rm TV}(\mathbf{X}) = \sum_{i=1}^{m}\sum_{j=1}^{n}\left(\left(\mathbf{X}\,\mathbf{D}_{1,n}^T\right)_{ij}^2 + \left(\mathbf{D}_{1,m}\,\mathbf{X}\right)_{ij}^2\right)^{1/2},$$

which is a sum of the 2-norms of all the gradients associated with the pixels in the image **X**. For periodic boundary conditions the circulant matrix $\mathbf{D}_{1,m}$ can be used to compute first derivatives, and for reflexive boundary conditions we suggest the choice

$$\mathbf{D}_{1,m} = \begin{bmatrix} -1 & 1 & & & \\ -1 & 0 & 1 & & \\ & \ddots & \ddots & \ddots & \\ & & -1 & 0 & 1 \\ & & & -1 & 1 \end{bmatrix},$$

which ensures that $J_{\rm TV}(\mathbf{X}) = \mathbf{0}$ when **X** is the constant image.

The total variation deblurring problem then takes the form of a nonlinear minimization problem

$$\min\left\{\|\mathbf{b} - \mathbf{A}\,\mathbf{x}\|_2^2 + \alpha\,J_{\rm TV}(\mathbf{X})\right\},$$

where the last, nonlinear term $J_{\rm TV}(\mathbf{X})$ plays the role of the regularization term from the Tikhonov formulation. This minimization problem can be solved by variations of Newton's method.

7.6 Blind Deconvolution

So far we have assumed that the PSF—and therefore the matrix **A**—is known exactly. In some cases, this is not a good assumption; for example, in Section 3.2 we discussed how the PSF might be measured in an experiment. If there is noise/errors in **A** as well as **b**, then this should be included in our model. Instead of the model

$$\mathbf{b} = \mathbf{A}\mathbf{x} + \mathbf{e},$$

we might use

$$\mathbf{b} = (\mathbf{A} + \mathbf{E}_A)\mathbf{x} + \mathbf{e},$$

where $\mathbf{E}_A$ and **e** are both unknown. This model has been given the rather unfortunate name of blind deconvolution. It arises, e.g., when we have an imprecise model for the deblurring, resulting in an imprecise PSF array **P** and therefore an imprecise blurring matrix **A**.

We can generalize our TSVD and Tikhonov solution methods to this more general model. For Tikhonov, instead of solving the regularized least squares minimization problem

$$\min_{\mathbf{x}}\left\{\|\mathbf{b} - \mathbf{A}\mathbf{x}\|_2^2 + \alpha^2\|\mathbf{D}\mathbf{x}\|_2^2\right\}$$

POINTER. The classic reference on total least squares problems is a book by Van Huffel and Vandewalle [58]. A discussion of regularization and applications to deblurring can be found in [14, 16]. Fast algorithms for structured problems are considered in [39, 48, 50]. Some blind deconvolution models avoid total least squares by assuming a special parametric form for the PSF; see, for example, [6].

over all choices of $\mathbf{x}$, we might solve

$$\min_{\mathbf{x}} \left\{ \|\mathbf{b} - (\mathbf{A} + \mathbf{E}_A)\mathbf{x}\|_2^2 + \|\mathbf{E}_A\|_F^2 + \alpha^2 \|\mathbf{D}\mathbf{x}\|_2^2 \right\}$$

over all choices of $\mathbf{x}$ and $\mathbf{E}_A$, where the Frobenius norm $\|\mathbf{E}_A\|_F$ is computed as the square root of the sum of the squares of the elements. This is known as a regularized total least squares problem.

If we assume that $\mathbf{E}_A$ has the same structure as $\mathbf{A}$ (e.g., BTTB), then to our generalized model we add this constraint, reducing the number of unknowns and speeding up the computation but making the solution algorithm more complicated.

7.7 When Spectral Methods Cannot Be Applied

We know that our deblurring problem can be solved by a fast spectral algorithm when the matrix $\mathbf{A}$ has certain special structure, as shown in the table in VIP 13. But what do we do when our problem does not fit into one of these frameworks?

For example, suppose that we have zero boundary conditions but the PSF array is not rank-one. Our matrix $\mathbf{A}$ is then BTTB, and our fast deblurring algorithms cannot be used. We do have two very nice properties, though. First, as mentioned in VIP 19, we can form products of $\mathbf{A}$ with arbitrary vectors quite fast. Second, the matrix is closely related to a BCCB matrix or a Kronecker product approximation, and deblurring with these matrices can be done quickly. How can we exploit these properties?

We can apply an iterative method to our deblurring problem, using the related BCCB or Kronecker product matrix as a preconditioner. In such a method, the main work per iteration is multiplication of a vector by $\mathbf{A}$ and solution of a linear system involving the preconditioner, and thus we exploit structure in the matrix $\mathbf{A}$. A good choice of iterative method is the LSQR algorithm of Paige and Saunders [47]. This method belongs to the family of Krylov subspace methods and is particularly well adapted to problems that need regularization.

In LSQR, we construct a sequence of approximate solutions $\mathbf{x}^{(j)}$ to the linear system $\mathbf{A}\mathbf{x} = \mathbf{b}$. After $N = mn$ steps, we have the (noise contaminated) naïve solution $\mathbf{x} = \mathbf{A}^{-1}\mathbf{b}$, so we will stop the iteration early. At iteration j, we solve the minimization problem

$$\min \|\mathbf{b} - \mathbf{A}\mathbf{x}\|_2^2$$

POINTER. A good reference on iterative methods in the Krylov subspace family and on preconditioning is the book by Saad [51]. LSQR is an algorithm due to Paige and Saunders [47].

over all vectors $\mathbf{x}$ in a j-dimensional subspace of the space of N-dimensional vectors. In order to produce a good regularized solution, we have several tools:

- The early subspaces generated by LSQR tend to contain spectral directions corresponding to the large singular values of $\mathbf{A}$. Therefore, we may be able to stop the iteration with a small value of j before the computed solution has significant components in directions corresponding to small singular values. In this case, the iterative method is the regularization method. We can use the discrepancy principle, L-curve, or GCV to choose the best value of j.

- We can choose a preconditioner to bias the subspaces so that this regularizing effect of the iteration is enhanced. This allows us to use a larger value of j without significant contamination by directions corresponding to small singular values and can yield a better deblurred image.

- Since j will remain small, we can afford to use our SVD-based methods on the $(j+1)\times j$ matrix generated by LSQR and use either Tikhonov or TSVD regularization. These methods are called hybrid methods for regularization; see, e.g., [36, 46] for details.

- Alternatively, we can apply LSQR to the Tikhonov-regularized problem

$$\min \left\| \begin{bmatrix} \mathbf{b} \\ \mathbf{0} \end{bmatrix} - \begin{bmatrix} \mathbf{A} \\ \alpha \mathbf{D} \end{bmatrix} \mathbf{x} \right\|_2^2 .$$

If LSQR is applied to the Tikhonov-regularized problem, then we need an approximation to $\mathbf{A}^T\mathbf{A} + \alpha^2\mathbf{D}^T\mathbf{D}$ as a preconditioner. For example, by replacing zero boundary conditions with periodic boundary conditions, we can construct BCCB approximations $\mathbf{A} \approx \mathbf{F}^*\mathbf{\Lambda}_A\mathbf{F}$ and $\mathbf{D} \approx \mathbf{F}^*\mathbf{\Lambda}_D\mathbf{F}$. Since

$$\begin{aligned} \mathbf{A}^T\mathbf{A} + \alpha^2\mathbf{D}^T\mathbf{D} &\approx (\mathbf{F}^*\mathbf{\Lambda}_A^*\mathbf{F})(\mathbf{F}^*\mathbf{\Lambda}_A\mathbf{F}) + \alpha^2(\mathbf{F}^*\mathbf{\Lambda}_D^*\mathbf{F})(\mathbf{F}^*\mathbf{\Lambda}_D\mathbf{F}) \\ &= \mathbf{F}^*(|\mathbf{\Lambda}_A|^2 + \alpha^2|\mathbf{\Lambda}_D|^2)\mathbf{F} \\ &\equiv \mathbf{M}^*\mathbf{M}, \end{aligned}$$

we might choose $\mathbf{M} = \mathbf{F}^*(|\mathbf{\Lambda}_A|^2 + \alpha^2|\mathbf{\Lambda}_D|^2)^{1/2}\mathbf{F}$ as our preconditioner. Note that for this preconditioner, regularization is built in, so $\mathbf{M}$ is well conditioned. We can use a similar approach for BTTB + BTHB + BHTB + BHHB approximations, using the DCT in place of the FFT.

If LSQR is applied to the minimization problem without regularization, then $\mathbf{M}$ must be chosen even more carefully. The danger is that if the preconditioner magnifies the noise, then our computed image will be useless. Therefore, if

$$\mathbf{A}^T\mathbf{A} \approx \mathbf{V}\widehat{\mathbf{\Sigma}}^2\mathbf{V}^T,$$

then we might use $\mathbf{M} = \mathbf{V}\mathbf{\Sigma}\mathbf{V}^T$ as a preconditioner, in which $\mathbf{\Sigma}$ has ones in place of the small spectral values of $\widehat{\mathbf{\Sigma}}$ to avoid magnifying error. We summarize the choices of preconditioners in Table 7.4

Table 7.4. *A summary of some choices of preconditioners for structured deblurring problems.*

PSF	Boundary condition	Matrix structure	Preconditioner
Rank > 1	Zero	BTTB	BCCB or Kronecker approximation
Nonsymmetric or rank > 1	Reflexive	BTTB + BTHB + BHTB + BHHB	Symmetric or Kronecker approximation
Spatially variant	Arbitrary	None	Any of the above

POINTER. We have outlined in this book just a small fraction of algorithms for deblurring images, and we have chosen them based on our own experience. Since the book is meant to be a tutorial rather than a research monograph, the references we give are also biased toward our own writing. We hope that what you have learned encourages you to investigate other classes of algorithms and to explore alternate viewpoints given in the work of the many other researchers in this area.

Appendix
MATLAB Functions

This three-part appendix includes MATLAB codes that illustrate how some of the different techniques and methods discussed in this book can be implemented. These codes, and the image data used in this book, may be obtained from the book's website. We emphasize that this is not intended to be a library or a complete software package; for more complete MATLAB packages, we suggest *Regularization Tools* [21] and *MOORe Tools* [30], which may be used for the analysis and solution of discrete ill-posed problems, and *Restore Tools* [43], which is an object-based package for iterative image deblurring algorithms.

1. TSVD Regularization Methods

Periodic Boundary Conditions

```
function [X, tol] = tsvd_fft(B, PSF, center, tol)
%TSVD_FFT Truncated SVD image deblurring using the FFT algorithm.
%
%function [X, tol] = tsvd_fft(B, PSF, center, tol)
%
%            X = tsvd_fft(B, PSF, center);
%            X = tsvd_fft(B, PSF, center, tol);
%     [X, tol] = tsvd_fft(B, PSF, ...);
%
%  Compute restoration using an FFT-based truncated spectral factorization.
%
%  Input:
%        B  Array containing blurred image.
%      PSF  Array containing the point spread function; same size as B.
%   center  [row, col] = indices of center of PSF.
%      tol  Regularization parameter (truncation tolerance).
%             Default parameter chosen by generalized cross validation.
%
%  Output:
%        X  Array containing computed restoration.
%      tol  Regularization parameter used to construct restoration.

% Reference: See Chapter 6,
```

```
%            "Deblurring Images - Matrices, Spectra, and Filtering"
%            by P. C. Hansen, J. G. Nagy, and D. P. O'Leary,
%            SIAM, Philadelphia, 2006.

%
% Check number of inputs and set default parameters.
%
if (nargin < 3)
   error('B, PSF, and center must be given.')
end
if (nargin < 4)
   tol = [];
end

%
% Use the FFT to compute the eigenvalues of the BCCB blurring matrix.
%
S = fft2( circshift(PSF, 1-center) );

%
% If a regularization parameter is not given, use GCV to find one.
%
bhat = fft2(B);
if (ischar(tol) | isempty(tol))
  tol = gcv_tsvd(S(:), bhat(:));
end

%
% Compute the TSVD regularized solution.
%
Phi = (abs(S) >= tol);
idx = (Phi~=0);
Sfilt = zeros(size(Phi));
Sfilt(idx) = Phi(idx) ./ S(idx);
X = real(ifft2(bhat .* Sfilt));
```

Reflexive Boundary Conditions

```
function [X, tol] = tsvd_dct(B, PSF, center, tol)

%TSVD_DCT Truncated SVD image deblurring using the DCT algorithm.
%
%function [X, tol] = tsvd_dct(B, PSF, center, tol)
%
%          X = tsvd_dct(B, PSF, center);
%          X = tsvd_dct(B, PSF, center, tol);
%     [X, tol] = tsvd_dct(B, PSF, ...);
%
%  Compute restoration using a DCT-based truncated spectral factorization.
%
%  Input:
%        B  Array containing blurred image.
%      PSF  Array containing the point spread function; same size as B.
%   center  [row, col] = indices of center of PSF.
%      tol  Regularization parameter (truncation tolerance).
%              Default parameter chosen by generalized cross validation.
%
```

```
% Output:
%        X  Array containing computed restoration.
%      tol  Regularization parameter used to construct restoration.

% Reference: See Chapter 6,
%            "Deblurring Images - Matrices, Spectra, and Filtering"
%            by P. C. Hansen, J. G. Nagy, and D. P. O'Leary,
%            SIAM, Philadelphia, 2006.

%
% Check number of inputs and set default parameters.
%
if (nargin < 3)
   error('B, PSF, and center must be given.')
end
if (nargin < 4)
   tol = [];
end

%
% Use the DCT to compute the eigenvalues of the symmetric
% BTTB + BTHB + BHTB + BHHB blurring matrix.
%
e1 = zeros(size(PSF)); e1(1,1) = 1;
% Check to see if the built-in dct2 function is available; if not,
% use our simple codes.
if exist('dct2') == 2
  bhat = dct2(B);
  S = dct2( dctshift(PSF, center) ) ./ dct2(e1);
else
  bhat = dcts2(B);
  S = dcts2( dctshift(PSF, center) ) ./ dcts2(e1);
end

%
% If a regularization parameter is not given, use GCV to find one.
%
if (ischar(tol) | isempty(tol))
  tol = gcv_tsvd(S(:), bhat(:));
end

%
% Compute the TSVD regularized solution.
%
Phi = (abs(S) >= tol);
idx = (Phi~=0);
Sfilt = zeros(size(Phi));
Sfilt(idx) = Phi(idx) ./ S(idx);
% Check again to see if the built-in dct2 function is available.
if exist('dct2') == 2
  X = idct2(bhat .* Sfilt);
else
  X = idcts2(bhat .* Sfilt);
end
```

Separable Two-Dimensional Blur

```
function [X, tol] = tsvd_sep(B, PSF, center, tol, BC)

%TSVD_SEP Truncated SVD image deblurring using Kronecker decomposition.
%
%function [X, tol] = tsvd_sep(B, PSF, center, tol, BC)
%
%          X = tsvd_sep(B, PSF, center);
%          X = tsvd_sep(B, PSF, center, tol);
%          X = tsvd_sep(B, PSF, center, tol, BC);
%     [X, tol] = tsvd_sep(B, PSF, ...);
%
%  Compute restoration using a Kronecker product decomposition and
%  a truncated SVD.
%
%  Input:
%        B  Array containing blurred image.
%      PSF  Array containing the point spread function; same size as B.
%   center  [row, col] = indices of center of PSF.
%      tol  Regularization parameter (truncation tolerance).
%             Default parameter chosen by generalized cross validation.
%       BC  String indicating boundary condition.
%             ('zero', 'reflexive', or 'periodic'; default is 'zero'.)
%
%  Output:
%        X  Array containing computed restoration.
%      tol  Regularization parameter used to construct restoration.

% Reference: See Chapter 6,
%            "Deblurring Images - Matrices, Spectra, and Filtering"
%            by P. C. Hansen, J. G. Nagy, and D. P. O'Leary,
%            SIAM, Philadelphia, 2006.

%
% Check number of inputs and set default parameters.
%
if (nargin < 3)
   error('B, PSF, and center must be given.')
end
if (nargin < 4)
   tol = [];
end
if (nargin < 5)
   BC = 'zero';
end

%
% First compute the Kronecker product terms, Ar and Ac, where
% A = kron(Ar, Ac).  Note that if the PSF is not separable, this
% step computes a Kronecker product approximation to A.
%
[Ar, Ac] = kronDecomp(PSF, center, BC);

%
% Compute SVD of the blurring matrix.
%
[Ur, Sr, Vr] = svd(Ar);
```

```
[Uc, Sc, Vc] = svd(Ac);

%
% If a regularization parameter is not given, use GCV to find one.
%
bhat = Uc'*B*Ur;
bhat = bhat(:);
s = kron(diag(Sr),diag(Sc));
if (ischar(tol) | isempty(tol))
  tol = gcv_tsvd(s, bhat(:));
end

%
% Compute the TSVD regularized solution.
%
Phi = (abs(s) >= tol);
idx = (Phi~=0);
Sfilt = zeros(size(Phi));
Sfilt(idx) = Phi(idx) ./ s(idx);
Bhat = reshape(bhat .*Sfilt , size(B));
X = Vc*Bhat*Vr';
```

Choosing Regularization Parameters

```
function tol = gcv_tsvd(s, bhat)

%GCV_TSVD Choose GCV parameter for TSVD image deblurring.
%
%function tol = gcv_tsvd(s, bhat)
%
%         tol = gcv_tsvd(s, bhat);
%
%  This function uses generalized cross validation (GCV) to choose
%  a truncation parameter for TSVD regularization.
%
%  Input:
%        s  Vector containing singular or spectral values.
%     bhat  Vector containing the spectral coefficients of the blurred
%             image.
%
%  Output:
%      tol  Truncation parameter; all abs(s) < tol should be truncated.

% Reference: See Chapter 6,
%            "Deblurring Images - Matrices, Spectra, and Filtering"
%            by P. C. Hansen, J. G. Nagy, and D. P. O'Leary,
%            SIAM, Philadelphia, 2006.

%
% Sort absolute values of singular/spectral values in descending order.
%
[s, idx] = sort(abs(s)); s = flipud(s); idx = flipud(idx);
bhat = abs( bhat(idx) );
n = length(s);
%
% The GCV function G for TSVD has a finite set of possible values
% corresponding to the truncation levels.  It is computed using
```

```
% rho, a vector containing the squared 2-norm of the residual for
% all possible truncation parameters tol.
%
rho = zeros(n-1,1);
rho(n-1) = bhat(n)^2;
G = zeros(n-1,1);
G(n-1) = rho(n-1);
for k=n-2:-1:1
  rho(k) = rho(k+1) + bhat(k+1)^2;
  G(k) = rho(k)/(n - k)^2;
end
% Ensure that the parameter choice will not be fooled by pairs of
% equal singular values.
for k=1:n-2,
  if (s(k)==s(k+1))
     G(k) = inf;
  end
end
%
% Now find the minimum of the discrete GCV function.
%
[minG,reg_min] = min(G);
%
% reg_min is the truncation index, and tol is the truncation parameter.
% That is, any singular values < tol are truncated.
%
tol = s(reg_min(1));
```

2. Tikhonov Regularization Methods

Periodic Boundary Conditions

```
function [X, alpha] = tik_fft(B, PSF, center, alpha)

%TIK_FFT Tikhonov image deblurring using the FFT algorithm.
%
%function [X, alpha] = tik_fft(B, PSF, center, alpha)
%
%            X = tik_fft(B, PSF, center);
%            X = tik_fft(B, PSF, center, alpha);
%   [X, alpha] = tik_fft(B, PSF, ...);
%
%  Compute restoration using an FFT-based Tikhonov filter,
%  with the identity matrix as the regularization operator.
%
%  Input:
%        B  Array containing blurred image.
%      PSF  Array containing the point spread function; same size as B.
%   center  [row, col] = indices of center of PSF.
%    alpha  Regularization parameter.
%             Default parameter chosen by generalized cross validation.
%
%  Output:
%        X  Array containing computed restoration.
%    alpha  Regularization parameter used to construct restoration.

% Reference: See Chapter 6,
```

```
%           "Deblurring Images - Matrices, Spectra, and Filtering"
%           by P. C. Hansen, J. G. Nagy, and D. P. O'Leary,
%           SIAM, Philadelphia, 2006.

%
% Check number of inputs and set default parameters.
%
if (nargin < 3)
   error('B, PSF, and center must be given.')
end
if (nargin < 4)
   alpha = [];
end

%
% Use the FFT to compute the eigenvalues of the BCCB blurring matrix.
%
S = fft2( circshift(PSF, 1-center) );
s = S(:);

%
% If a regularization parameter is not given, use GCV to find one.
%
bhat = fft2(B);
bhat = bhat(:);
if (ischar(alpha) | isempty(alpha))
  alpha = gcv_tik(s, bhat);
end

%
% Compute the Tikhonov regularized solution.
%
D = conj(s).*s + abs(alpha)^2;
bhat = conj(s) .* bhat;
xhat = bhat ./ D;
xhat = reshape(xhat, size(B));
X = real(ifft2(xhat));
```

Reflexive Boundary Conditions

```
function [X, alpha] = tik_dct(B, PSF, center, alpha)

%TIK_DCT Tikhonov image deblurring using the DCT algorithm.
%
%function [X, alpha] = tik_dct(B, PSF, center, alpha)
%
%            X = tik_dct(B, PSF, center);
%            X = tik_dct(B, PSF, center, alpha);
%   [X, alpha] = tik_dct(B, PSF, ...);
%
%  Compute restoration using a DCT-based Tikhonov filter,
%  with the identity matrix as the regularization operator.
%
%  Input:
%        B  Array containing blurred image.
%      PSF  Array containing the point spread function; same size as B.
%   center  [row, col] = indices of center of PSF.
```

```
%    alpha  Regularization parameter.
%             Default parameter chosen by generalized cross validation.
%
%  Output:
%         X  Array containing computed restoration.
%    alpha  Regularization parameter used to construct restoration.

% Reference: See Chapter 6,
%            "Deblurring Images - Matrices, Spectra, and Filtering"
%            by P. C. Hansen, J. G. Nagy, and D. P. O'Leary,
%            SIAM, Philadelphia, 2006.

%
% Check number of inputs and set default parameters.
%
if (nargin < 3)
   error('B, PSF, and center must be given.')
end
if (nargin < 4)
   alpha = [];
end

%
% Use the DCT to compute the eigenvalues of the symmetric
% BTTB + BTHB + BHTB + BHHB blurring matrix.
%
% Check to see if the built-in dct2 function is available; if not,
% use our simple codes.
e1 = zeros(size(PSF)); e1(1,1) = 1;
if exist('dct2') == 2
  bhat = dct2(B);
  S = dct2( dctshift(PSF, center) ) ./ dct2(e1);
else
  bhat = dcts2(B);
  S = dcts2( dctshift(PSF, center) ) ./ dcts2(e1);
end

%
% If a regularization parameter is not given, use GCV to find one.
%
bhat = bhat(:);
s = S(:);
if (ischar(alpha) | isempty(alpha))
  alpha = gcv_tik(s, bhat);
end

%
% Compute the Tikhonov regularized solution.
%
D = conj(s).*s + abs(alpha)^2;
bhat = conj(s) .* bhat;
xhat = bhat ./ D;
xhat = reshape(xhat, size(B));
% Check again to see if the built-in dct2 function is available.
if exist('dct2') == 2
  X = idct2(xhat);
```

```
else
  X = idcts2(xhat);
end
```

Separable Two-Dimensional Blur

```
function [X, alpha] = tik_sep(B, PSF, center, alpha, BC)

%TIK_SEP Tikhonov image deblurring using the Kronecker decomposition.
%
%function [X, alpha] = tik_sep(B, PSF, center, alpha, BC)
%
%            X = tik_sep(B, PSF, center);
%            X = tik_sep(B, PSF, center, alpha);
%            X = tik_sep(B, PSF, center, alpha, BC);
%   [X, alpha] = tik_sep(B, PSF, ...);
%
%  Compute restoration using a Kronecker product decomposition and a
%  Tikhonov filter, with the identity matrix as the regularization operator.
%
%  Input:
%        B  Array containing blurred image.
%      PSF  Array containing the point spread function; same size as B.
%   center  [row, col] = indices of center of PSF.
%    alpha  Regularization parameter.
%             Default parameter chosen by generalized cross validation.
%       BC  String indicating boundary condition.
%             ('zero', 'reflexive', or 'periodic')
%           Default is 'zero'.
%
%  Output:
%        X  Array containing computed restoration.
%    alpha  Regularization parameter used to construct restoration.

% Reference: See Chapter 6,
%            "Deblurring Images - Matrices, Spectra, and Filtering"
%            by P. C. Hansen, J. G. Nagy, and D. P. O'Leary,
%            SIAM, Philadelphia, 2006.

%
% Check number of inputs and set default parameters.
%
if (nargin < 3)
   error('B, PSF, and center must be given.')
end
if (nargin < 4)
   alpha = [];
end
if (nargin < 5)
   BC = 'zero';
end

%
% First compute the Kronecker product terms, Ar and Ac, where
% the blurring matrix  A = kron(Ar, Ac).
% Note that if the PSF is not separable, this
```

```
% step computes a Kronecker product approximation to A.
%
[Ar, Ac] = kronDecomp(PSF, center, BC);

%
% Compute SVD of the blurring matrix.
%
[Ur, Sr, Vr] = svd(Ar);
[Uc, Sc, Vc] = svd(Ac);

%
% If a regularization parameter is not given, use GCV to find one.
%
bhat = Uc'*B*Ur;
bhat = bhat(:);
s = kron(diag(Sr),diag(Sc));
if (ischar(alpha) | isempty(alpha))
  alpha = gcv_tik(s, bhat);
end

%
% Compute the Tikhonov regularized solution.
%
D = abs(s).^2 + abs(alpha)^2;
bhat = s .* bhat;
xhat = bhat ./ D;
xhat = reshape(xhat, size(B));
X = Vc*xhat*Vr';
```

Choosing Regularization Parameters

```
function alpha = gcv_tik(s, bhat)

%GCV_TIK Choose GCV parameter for Tikhonov image deblurring.
%
%function alpha = gcv_tik(s, bhat)
%
%         alpha = gcv_tik(s, bhat);
%
%  This function uses generalized cross validation (GCV) to choose
%  a regularization parameter for Tikhonov filtering.
%
%  Input:
%        s  Vector containing singular or spectral values
%             of the blurring matrix.
%     bhat  Vector containing the spectral coefficients of the blurred
%             image.
%
%  Output:
%    alpha  Regularization parameter.

% Reference: See Chapter 6,
%            "Deblurring Images - Matrices, Spectra, and Filtering"
%            by P. C. Hansen, J. G. Nagy, and D. P. O'Leary,
%            SIAM, Philadelphia, 2006.
alpha = fminbnd(@GCV, min(abs(s)), max(abs(s)), [], s, bhat);
```

```
  function G = GCV(alpha, s, bhat)
    %
    %  This is a nested function that evaluates the GCV function for
    %  Tikhonov filtering.  It is called by fminbnd.
    %
    phi_d = 1 ./ (abs(s).^2 + alpha^2);
    G = sum(abs(bhat.*phi_d).^2) / (sum(phi_d)^2);
  end

end
```

3. Auxiliary Functions

```
function y = dcts(x)
%DCTS Model implementation of discrete cosine transform.
%
%function y = dcts(x)
%
%         y = dcts(x);
%
%  Compute the discrete cosine transform of x.
%  This is a very simple implementation.  If the Signal Processing
%  Toolbox is available, then you should use the function dct.
%
%  Input:
%        x  column vector, or a matrix.  If x is a matrix then dcts(x)
%           computes the DCT of each column
%
%  Output:
%        y  contains the discrete cosine transform of x.

% Reference: See Chapter 4,
%            "Deblurring Images - Matrices, Spectra, and Filtering"
%            by P. C. Hansen, J. G. Nagy, and D. P. O'Leary,
%            SIAM, Philadelphia, 2006.
%
% If an FFT routine is available, then it can be used to compute
% the DCT.  Since the FFT is part of the standard MATLAB distribution,
% we use this approach.  For further details on the formulas, see:


%
%            "Computational Frameworks for the Fast Fourier Transform"
%            by C. F. Van Loan, SIAM, Philadelphia, 1992.
%
%            "Fundamentals of Digital Image Processing"
%            by A. Jain, Prentice-Hall, NJ, 1989.
%
[n, m] = size(x);

omega = exp(-i*pi/(2*n));
d = [1/sqrt(2); omega.^(1:n-1).'] / sqrt(2*n);
d = d(:,ones(1,m));


xt = [x; flipud(x)];
```

```
yt = fft(xt);
y = real(d .* yt(1:n,:));
```

```
function y = dcts2(x)

%DCTS2 Model implementation of 2-D discrete cosine transform.
%
%function y = dcts2(x)
%
%         y = dcts2(x);
%
%  Compute the two-dimensional discrete cosine transform of x.
%  This is a very simple implementation.  If the Image Processing Toolbox
%  is available, then you should use the function dct2.
%
%  Input:
%        x  array
%
%  Output:
%        y  contains the two-dimensional discrete cosine transform of x.

% Reference: See Chapter 4,
%            "Deblurring Images - Matrices, Spectra, and Filtering"
%            by P. C. Hansen, J. G. Nagy, and D. P. O'Leary,
%            SIAM, Philadelphia, 2006.
%
%            See also:
%            "Computational Frameworks for the Fast Fourier Transform"
%            by C. F. Van Loan, SIAM, Philadelphia, 1992.
%
%            "Fundamentals of Digital Image Processing"
%            by A. Jain, Prentice Hall, NJ, 1989.
%
% The two-dimensional DCT is obtained by computing a one-dimensional DCT of
% the columns, followed by a one-dimensional DCT of the rows.
%
y = dcts(dcts(x).').';
```

```
function Ps = dctshift(PSF, center)

%DCTSHIFT Create array containing the first column of a blurring matrix.
%
%function Ps = dctshift(PSF, center)
%
%         Ps = dctshift(PSF, center);
%
%  Create an array containing the first column of a blurring matrix
%  when implementing reflexive boundary conditions.
%
%  Input:
%      PSF  Array containing the point spread function.
%   center  [row, col] = indices of center of PSF.
%
%  Output:
%       Ps  Array (vector) containing first column of blurring matrix.
```

```
% Reference: See Chapter 4,
%            "Deblurring Images - Matrices, Spectra, and Filtering"
%            by P. C. Hansen, J. G. Nagy, and D. P. O'Leary,
%            SIAM, Philadelphia, 2006.

[m,n] = size(PSF);

if nargin == 1
  error('The center must be given.')
end

i = center(1);
j = center(2);
k = min([i-1,m-i,j-1,n-j]);

%
% The PSF gives the entries of a central column of the blurring matrix.
% The first column is obtained by reordering the entries of the PSF; for
% a detailed description of this reordering, see the reference cited
% above.
%
PP = PSF(i-k:i+k,j-k:j+k);

Z1 = diag(ones(k+1,1),k);
Z2 = diag(ones(k,1),k+1);

PP = Z1*PP*Z1' + Z1*PP*Z2' + Z2*PP*Z1' + Z2*PP*Z2';

Ps = zeros(m,n);
Ps(1:2*k+1,1:2*k+1) = PP;
```

```
function y = idcts(x)

%IDCTS Model implementation of inverse discrete cosine transform.
%
%function y = idcts(x)
%
%         y = idcts(x);
%
%  Compute the inverse discrete cosine transform of x.
%  This is a very simple implementation.  If the Signal Processing
%  Toolbox is available, then you should use the function idct.
%
%  Input:
%        x  column vector, or a matrix.  If x is a matrix then idcts
%           computes the IDCT of each column.
%
%  Output:
%        y  contains the inverse discrete cosine transform of x.

% Reference: See Chapter 4,
%            "Deblurring Images - Matrices, Spectra, and Filtering"
%            by P. C. Hansen, J. G. Nagy, and D. P. O'Leary,
%            SIAM, Philadelphia, 2006.
%
%
```

```
% If an inverse FFT routine is available, then it can be used to compute
% the inverse DCT.  Since the inverse FFT is part of the standard MATLAB
% distribution, we use this approach.  For further details on the formulas,
% see
%            "Computational Frameworks for the Fast Fourier Transform"
%            by C. F. Van Loan, SIAM, Philadelphia, 1992.
%
%            "Fundamentals of Digital Image Processing"
%            by A. Jain, Prentice-Hall, NJ, 1989.
%
[n, m] = size(x);

omega = exp(i*pi/(2*n));
d = sqrt(2*n) * omega.^(0:n-1).';
d(1) = d(1) * sqrt(2);
d = d(:,ones(1,m));

xt = [d.*x; zeros(1,m); -i*d(2:n,:).*flipud(x(2:n,:))];
yt = ifft(xt);
y = real(yt(1:n,:));
```

```
function y = idcts2(x)

%IDCTS2 Model implementation of 2-D inverse discrete cosine transform.
%
%function y = idcts2(x)
%
%        y = idcts2(x);
%
%  Compute the inverse two-dimensional discrete cosine transform of x.
%  This is a very simple implementation.  If the Image Processing Toolbox
%  is available, then you should use the function idct2.
%
%  Input:
%       x  array
%
%  Output:
%       y  contains the two-dimensional inverse discrete cosine
%          transform of x.

% Reference: See Chapter 4,
%            "Deblurring Images - Matrices, Spectra, and Filtering"
%            by P. C. Hansen, J. G. Nagy, and D. P. O'Leary,
%            SIAM, Philadelphia, 2006.
%
%            See also:
%            "Computational Frameworks for the Fast Fourier Transform"
%            by C. F. Van Loan, SIAM, Philadelphia, 1992.
%
%            "Fundamentals of Digital Image Processing"
%            by A. Jain, Prentice-Hall, NJ, 1989.
%
% The two-dimensional inverse DCT is obtained by computing a one-dimensional
% inverse DCT of the columns, followed by a one-dimensional inverse DCT of
% the rows.
```

```
%
y = idcts(idcts(x).').';
```

```
function [Ar, Ac] = kronDecomp(P, center, BC)

%KRONDECOMP Kronecker product decomposition of a PSF array
%
%function [Ar, Ac] = kronDecomp(P, center, BC)
%
%      [Ar, Ac] = kronDecomp(P, center);
%      [Ar, Ac] = kronDecomp(P, center, BC);
%
%  Compute terms of Kronecker product factorization A = kron(Ar, Ac),
%  where A is a blurring matrix defined by a PSF array.  The result is
%  an approximation only, if the PSF array is not rank-one.
%
%  Input:
%        P  Array containing the point spread function.
%   center  [row, col] = indices of center of PSF, P.
%       BC  String indicating boundary condition.
%             ('zero', 'reflexive', or 'periodic')
%           Default is 'zero'.
%
%  Output:
%   Ac, Ar  Matrices in the Kronecker product decomposition.  Some notes:
%               * If the PSF, P is not separable, a warning is displayed
%                 indicating the decomposition is only an approximation.
%               * The structure of Ac and Ar depends on the BC:
%                   zero      ==> Toeplitz
%                   reflexive ==> Toeplitz-plus-Hankel
%                   periodic  ==> circulant

% Reference: See Chapter 4,
%            "Deblurring Images - Matrices, Spectra, and Filtering"
%            by P. C. Hansen, J. G. Nagy, and D. P. O'Leary,
%            SIAM, Philadelphia, 2006.

%
% Check inputs and set default parameters.
%
if (nargin < 2)
   error('P and center must be given.')
end
if (nargin < 3)
   BC = 'zero';
end

%
% Find the two largest singular values and corresponding singular vectors
% of the PSF -- these are used to see if the PSF is separable.
%
[U, S, V] = svds(P, 2);
if ( S(2,2) / S(1,1) > sqrt(eps) )
  warning('The PSF, P is not separable; using separable approximation.')
end

%
```

```
% Since the PSF has nonnegative entries, we would like the vectors of the
% rank-one decomposition of the PSF to have nonnegative components.  That
% is, the singular vectors corresponding to the largest singular value of
% P should have nonnegative entries. The next few statements check this,
% and change sign if necessary.
%
minU = abs(min(U(:,1)));
maxU = max(abs(U(:,1)));
if minU == maxU
  U = -U;
  V = -V;
end

%
% The matrices Ar and Ac are defined by vectors r and c, respectively.
% These vectors can be computed as follows:
%
c = sqrt(S(1,1))*U(:,1);
r = sqrt(S(1,1))*V(:,1);

%
% The structure of Ar and Ac depends on the imposed boundary condition.
%
switch BC
  case 'zero'
    % Build Toeplitz matrices here
    Ar = buildToep(r, center(2));
    Ac = buildToep(c, center(1));
  case 'reflexive'
    % Build Toeplitz-plus-Hankel matrices here
    Ar = buildToep(r, center(2)) + buildHank(r, center(2));
    Ac = buildToep(c, center(1)) + buildHank(c, center(1));
  case 'periodic'
    % Build circulant matrices here
    Ar = buildCirc(r, center(2));
    Ac = buildCirc(c, center(1));
  otherwise
    error('Invalid boundary condition.')
end



function T = buildToep(c, k)
%
%  Build a banded Toeplitz matrix from a central column and an index
%  denoting the central column.
%
n = length(c);
col = zeros(n,1);
row = col';
col(1:n-k+1,1) = c(k:n);
row(1,1:k) = c(k:-1:1)';
T = toeplitz(col, row);

function C = buildCirc(c, k)
%
%  Build a banded circulant matrix from a central column and an index
```

```
%  denoting the central column.
%
n = length(c);
col = [c(k:n); c(1:k-1)];
row = [c(k:-1:1)', c(n:-1:k+1)'];
C = toeplitz(col, row);


function H = buildHank(c, k)
%
%  Build a Hankel matrix for separable PSF and reflexive boundary
%  conditions.
%
n = length(c);
col = zeros(n,1);
col(1:n-k) = c(k+1:n);
row = zeros(n,1);
row(n-k+2:n) = c(1:k-1);
H = hankel(col, row);
```

```
function P = padPSF(PSF, m, n)

%PADPSF Pad a PSF array with zeros to make it bigger.
%
%function P = padPSF(PSF, m, n)
%
%       P = padPSF(PSF, m);
%       P = padPSF(PSF, m, n);
%       P = padPSF(PSF, [m,n]);
%
%  Pad PSF with zeros to make it an m-by-n array.
%
%  If the PSF is an array with dimension smaller than the blurred image,
%  then deblurring codes may require padding first, such as:
%      PSF = padPSF(PSF, size(B));
%  where B is the blurred image array.
%
%  Input:
%      PSF  Array containing the point spread function.
%     m, n  Desired dimension of padded array.
%             If only m is specified, and m is a scalar, then n = m.
%
%  Output:
%        P  Padded m-by-n array.

% Reference: See Chapter 4,
%            "Deblurring Images - Matrices, Spectra, and Filtering"
%            by P. C. Hansen, J. G. Nagy, and D. P. O'Leary,
%            SIAM, Philadelphia, 2006.

%
% Set default parameters.
%
if nargin == 2
  if length(m) == 1
    n = m;
```

```
  else
    n = m(2);
    m = m(1);
  end
end

%
% Pad the PSF with zeros.
%
P = zeros(m, n);
P(1:size(PSF,1), 1:size(PSF,2)) = PSF;
```

Bibliography

[1] H. Andrews and B. Hunt. *Digital Image Restoration*. Prentice–Hall, Englewood Cliffs, NJ, 1977. (Cited on pp. 8, 22.)

[2] J. M. Bardsley and C. R. Vogel. A nonnegatively constrained convex programming method for image reconstruction. *SIAM J. Sci. Comput.*, 25:1326–1343, 2003. (Cited on pp. 28, 29.)

[3] M. Bertero and P. Boccacci. *Introduction to Inverse Problems in Imaging*. IOP Publishing Ltd., London, 1998. (Cited on p. 22.)

[4] Å. Björck. *Numerical Methods for Least Squares Problems*. SIAM, Philadelphia, 1996. (Cited on pp. 9, 91.)

[5] P. Blomgren and T. F. Chan. Modular solvers for constrained image restoration problems using the discrepancy principle. *Numer. Linear Algebra Appl.*, 9:347–358, 2002. (Cited on p. 80.)

[6] A. S. Carasso. Direct blind deconvolution. *SIAM J. Appl. Math.*, 61:1980–2007, 2001. (Cited on p. 100.)

[7] T. F. Chan and J. Shen. *Image Processing and Analysis: Variational, PDE, Wavelet, and Stochastic Methods*. SIAM, Philadelphia, 2005. (Cited on p. 99.)

[8] P. J. Davis. *Circulant Matrices*. Wiley, New York, 1979. (Cited on p. 42.)

[9] M. H. DeGroot. *Probability and Statistics*. Addision–Wesley, Reading, MA, 1989. (Cited on p. 84.)

[10] L. Desbat and D. Girard. The "minimum reconstruction error" choice of regularization parameters: Some more efficient methods and their application to deconvolution problems. *SIAM J. Sci. Comput.*, 16:1387–1403, 1995. (Cited on pp. 80, 81.)

[11] A. Diaspro, M. Corosu, P. Ramoino, and M. Robello. Two-photon excitation imaging based on a compact scanning head. *IEEE Engineering in Medicine and Biology*, pages 18–22, September/October 1999. (Cited on p. 23.)

[12] H. W. Engl and W. Grever. Using the L-curve for determining optimal regularization parameters. *Numer. Math.*, 69:25–31, 1994. (Cited on p. 82.)

[13] S. Farsiu, D. Robinson, M. Elad, and P. Milanfar. Advances and challenges in super-resolution. *Int. J. Imaging Syst. Technol.*, 14(2):47–57, 2004. (Cited on p. 96.)

[14] R. D. Fierro, G. H. Golub, P. C. Hansen, and D. P. O'Leary. Regularization by truncated total least squares. *SIAM J. Sci. Comput.*, 18:1223–1241, 1997. (Cited on p. 100.)

[15] D. A. Forsyth and J. Ponce. *Computer Vision. A Modern Appraoch.* Prentice–Hall, Englewood Cliffs, NJ, 2003. (Cited on p. 22.)

[16] G. H. Golub, P. C. Hansen, and D. P. O'Leary. Tikhonov regularization and total least squares. *SIAM J. Matrix Anal. Appl.*, 21:185–194, 1999. (Cited on p. 100.)

[17] G. H. Golub, M. Heath, and G. Wahba. Generalized cross-validation as a method for choosing a good ridge parameter. *Technometrics*, 21:215–223, 1979. (Cited on p. 80.)

[18] G. H. Golub and C. F. Van Loan. *Matrix Computations, Third Edition.* Johns Hopkins University Press, Baltimore, 1996. (Cited on p. 9.)

[19] W. Groetsch. *Theory of Tikhonov Regularization for Fredholm Equations of the First Kind.* Pitman Publishing Ltd., Boston, MA, 1984. (Cited on p. 80.)

[20] P. C. Hansen. Analysis of discrete ill-posed problems by means of the L-curve. *SIAM Rev.*, 34:561–580, 1992. (Cited on p. 81.)

[21] P. C. Hansen. Regularization tools: A MATLAB package for the analysis and solution of discrete ill-posed problems. *Numer. Algorithms*, 6:1–35, 1994. See also http://www2.imm.dtu.dk/~pch/Regutools. (Cited on p. 103.)

[22] P. C. Hansen. *Rank-Deficient and Discrete Ill-Posed Problems.* SIAM, Philadelphia, 1998. (Cited on pp. 81, 91.)

[23] P. C. Hansen. Deconvolution and regularization with Toeplitz matrices. *Numer. Algorithms*, 29:323–378, 2002. (Cited on p. 8.)

[24] P. C. Hansen and D. P. O'Leary. The use of the L-curve in the regularization of discrete ill-posed problems. *SIAM J. Sci. Comput.*, 14:1487–1503, 1993. (Cited on p. 81.)

[25] J. W. Hardy. Adaptive optics. *Scientific American*, 270(6):60–65, 1994. (Cited on p. 23.)

[26] S. Haykin. *Adaptive Filter Theory.* Prentice–Hall, Englewood Cliffs, NJ, 1991. (Cited on p. 96.)

[27] D. J. Higham and N. J. Higham. *MATLAB Guide, Second Edition.* SIAM, Philadelphia, 2005. (Cited on p. 2.)

[28] N. J. Higham. *Accuracy and Stability of Numerical Algorithms, Second Edition.* SIAM, Philadelphia, 2002. (Cited on p. 91.)

[29] B. Hofmann. *Regularization for Applied Inverse and Ill-Posed Problems.* Teubner-Texte Mathe., Teubner, Leipzig, 1986. (Cited on p. 81.)

[30] M. Jacobsen. *Modular Regularization Algorithms*. Ph.D. thesis, Informatics and Mathematical Modelling, Technical University of Denmark, Lyngby, Denmark, 2004. See also http://www2.imm.dtu.dk/~pch/MOOReTools. (Cited on p. 103.)

[31] A. K. Jain. *Fundamentals of Digital Image Processing*. Prentice–Hall, Englewood Cliffs, NJ, 1989. (Cited on pp. 22, 25, 28, 42, 47.)

[32] N. L. Johnson, S. Kotz, and N. Balakrishnan. *Continuous Univariate Distributions, Vol.* 1; *Second Edition*. Wiley, New York, 1994. (Cited on p. 85.)

[33] L. Joyeux, S. Boukir, B. Besserer, and O. Buisson. Reconstruction of degraded image sequences. Application to film restoration. *Image and Vision Computing*, 19:504–516, 2000. (Cited on p. 1.)

[34] D. W. Kammler. *A First Course in Fourier Analysis*. Prentice–Hall, Englewood Cliffs, NJ, 2000. (Cited on p. 73.)

[35] N. B. Karayiannis and A. N. Venetsanopoulos. Regularization theory in image restoration—the stabilizing functional approach. *IEEE Trans. Acoustics, Speech, and Signal Processing*, 38:1155–1179, 1990. (Cited on p. 96.)

[36] M. E. Kilmer and D. P. O'Leary. Choosing regularization parameters in iterative methods for ill-posed problems. *SIAM J. Matrix Anal. Appl.*, 22:1204–1221, 2001. (Cited on pp. 81, 101.)

[37] R. L. Lagendijk and J. Biemond. *Iterative Identification and Restoration of Images*. Kluwer Academic Publishers, Boston/Dordrecht/London, 1991. (Cited on p. 22.)

[38] C. L. Lawson and R. J. Hanson. *Solving Least Squares Problems*. Prentice–Hall, Englewood Cliffs, NJ, 1974. Reprinted SIAM, Philadelphia, 1995. (Cited on p. 81.)

[39] N. Mastronardi, P. Lemmerling, and S. Van Huffel. Fast structured total least squares algorithm for solving the basic deconvolution problem. *SIAM J. Matrix Anal. Appl.*, 22:533–553, 2000. (Cited on p. 100.)

[40] N. Mastronardi and D. P. O'Leary. Robust regression and ℓ_1 approximations for Toeplitz problems. To appear. (Cited on p. 97.)

[41] A. F. J. Moffat. A theoretical investigation of focal stellar images in the photographic emulsion and applications to photographic photometry. *Astronom. Astrophys.*, 3:455–461, 1969. (Cited on p. 26.)

[42] V. A. Morozov. On the solution of functional equations by the method of regularization. *Soviet Math. Dokl.*, 7:414–417, 1966. (Cited on p. 80.)

[43] J. G. Nagy, K. M. Palmer, and L. Perrone. Iterative methods for image deblurring: A MATLAB object oriented approach. *Numer. Algorithms*, 36:73–93, 2004. See also http://www.mathcs.emory.edu/~nagy/RestoreTools. (Cited on p. 103.)

[44] M. K. Ng, R. H. Chan, and W.-C. Tang. A fast algorithm for deblurring models with Neumann boundary conditions. *SIAM J. Sci. Comput.*, 21:851–866, 1999. (Cited on p. 45.)

[45] D. P. O'Leary. Robust regression computation using iteratively reweighted least squares. *SIAM J. Matrix Anal. Appl.*, 11:466–480, 1990. (Cited on p. 97.)

[46] D. P. O'Leary and J. A. Simmons. A bidiagonalization-regularization procedure for large scale discretizations of ill-posed problems. *SIAM J. Sci. Stat. Comput.*, 2:474–489, 1981. (Cited on p. 101.)

[47] C. C. Paige and M. A. Saunders. LSQR: An algorithm for sparse linear equations and sparse least squares. *ACM Trans. Math. Software*, 8:43–71, 1982. (Cited on p. 100.)

[48] A. Pruessner and D. P. O'Leary. Blind deconvolution using a regularized structured total least norm approach. *SIAM J. Matrix Anal. Appl.*, 24:1018–1037, 2003. (Cited on pp. 97, 100.)

[49] M. C. Roggemann and B. Welsh. *Imaging Through Turbulence*. CRC Press, Boca Raton, FL, 1996. (Cited on pp. 22, 25.)

[50] J. B. Rosen, H. Park, and J. Glick. Total least norm formulation and solution for structured problems. *SIAM J. Matrix Anal. Appl.*, 17:110–126, 1996. (Cited on p. 100.)

[51] Y. Saad. *Iterative Methods for Sparse Linear Systems, Second Edition*. SIAM, Philadelphia, 2003. (Cited on p. 100.)

[52] K. S. Shanmugan and A. M. Breipohl. *Random Signals: Detection, Estimation and Data Analysis*. Wiley, New York, 1988. (Cited on p. 84.)

[53] G. Sharma and H. J. Trussell. Digital color imaging. *IEEE Trans. Image Process.*, 6:901–932, 1997. (Cited on p. 90.)

[54] S. W. Smith. *The Scientist and Engineer's Guide to Digital Signal Processing*. California Techincal Publishing, San Diego, CA, 1997. (Cited on p. 53.)

[55] D. L. Snyder, C. W. Hammoud, and R. L. White. Image recovery from data acquired with a charge-coupled-device camera. *J. Opt. Soc. Amer. A*, 10:1014–1023, 1993. (Cited on p. 28.)

[56] D. L. Snyder, C. W. Helstrom, and A. D. Lanterman. Compensation for readout noise in CCD images. *J. Opt. Soc. Amer. A*, 12:272–283, 1994. (Cited on p. 28.)

[57] G. W. Stewart. *Matrix Algorithms: Volume* 1: *Basic Decompositions*. SIAM, Philadelphia, 1998. (Cited on p. 9.)

[58] S. Van Huffel and J. Vandewalle. *The Total Least Squares Problem: Computational Aspects and Analysis*. SIAM, Philadelphia, 1991. (Cited on p. 100.)

[59] C. F. Van Loan. *Computational Frameworks for the Fast Fourier Transform*. SIAM, Philadelphia, 1992. (Cited on pp. 39, 42, 47.)

[60] J. M. Varah. Pitfalls in the numerical solution of linear ill-posed problems. *SIAM J. Sci. Stat. Comput.*, 4:164–176, 1983. (Cited on p. 82.)

[61] C. R. Vogel. Non-convergence of the L-curve regularization parameter selection method. *Inverse Problems*, 12:535–547, 1996. (Cited on p. 82.)

[62] C. R. Vogel. *Computational Methods for Inverse Problems*. SIAM, Philadelphia, 2002. (Cited on p. 99.)

[63] T. Wittman. Lost in the supermarket: Decoding blurry barcodes. *SIAM News*, 37(7):16, September 2004. (Cited on p. 5.)

Index

backslash, 6, 49
basis image, 61–64
Bayer pattern, 87
binary image, 13
blind deconvolution, 99
blur, 21–24, 27
  astigmatism, 22
  atmospheric turbulence, 1, 22, 25, 32, 45
  cross-channel, 88, 90
  Gaussian, 25–27, 32, 63, 67
  matrix, 8, 12
  model, 4, 7, 22, 23, 30
  Moffat, 26, 27
  motion, 22, 25, 27
  out-of-focus, 1, 4, 22, 25, 27
  separable, 38
  spatially invariant, 24, 27, 52, 89
  spatially variant, 102
  within-channel, 88
boundary condition, 24, 29–32, 53, 57, 58, 75
  periodic, 30, 36, 38–44, 52, 63, 75, 86, 92, 95, 96, 99
  reflexive, 31, 36, 38–40, 44–48, 52, 58, 63, 75, 76, 86, 99, 102
  zero, 30, 31, 36–40, 49, 57–59, 100, 102

charge-coupled device (CCD), 4, 22, 28, 29, 87
color
  CMY, 14
  CMYK, 14
  HSV, 14
  RGB, 13
color image, 2, 4, 13, 17, 87–90
condition number, 7, 10, 55, 59
conj, 33, 63
convolution
  one-dimensional, 34, 39
  two-dimensional, 27, 28, 35, 37
covariance matrix, 84

deblurring, 23
discrepancy principle, 80, 82, 85
discrete cosine transform (DCT), 45–47, 51, 52, 54, 63–66, 68, 69
discrete Picard condition, 67–70, 78, 79

edge detection, 28
edge enhancement, 28
eigenvalues, 33, 42–44, 46, 48
eigenvectors, 72
error, 1, 5, 10, 72, 85
  high frequency, 11
  perturbation, 10, 77–79, 81
  quantization, 6, 29
  random (*see* noise), 5
  regularization, 77–79, 81
  relative, 7
  rounding, 10

filter factor, 57, 59, 71–74, 78–80
  Tikhonov, 72–75
  TSVD, 72, 73, 75
filtering methods, 2, 71, 74
  high-pass, 28
  low-pass, 28, 69, 73
  pseudo-inverse, 56
  spectral, 55–57, 59, 61, 71–74, 77–80
finite difference, 92

Flexible Image Transport System (FITS), 20
forward slash, 49
Fourier transform, 27, 73
  coordinate system, 73
  discrete (DFT), 41, 42, 63–69
  fast (FFT), 41, 42, 47, 51, 52, 54

generalized cross validation (GCV), 80–83, 85, 101
GIF, 16
grayscale image, 4, 13–15, 17, 19, 23

high-contrast image, 20
hybrid method, 101

Image Processing Toolbox (IPT), 2, 14, 15, 17, 19, 29, 96
integral equation, 8
iterative method, 100

JPEG, 15–17, 19

Krylov subspace method, 100

L-curve, 81–82
Laplacian, 93–96
least squares, 91, 97
  damped, 90
  iteratively reweighted, 97
  total, 100
linear blur, 4, 7, 8, 21–23
LSQR, 100

MATLAB, 2
  axis, 2, 15, 19
  circshift, 34, 42
  colormap, 2, 13, 15, 16, 19
  cond, 7
  conv2, 21, 27, 69
  dct2, 34, 45, 47, 48, 67, 69
  dcts, 47, 113
  dcts2, 34, 47, 48, 114
  dctshift, 34, 47, 75, 114
  deconvreg, 96
  deconvwnr, 96
  diag, 47, 75, 83
  doc, 14
  double, 13, 17
  eig, 43
  fft2, 34, 41–44, 47, 66, 67, 69, 75
  figure, 15
  fliplr, 31, 45
  flipud, 31
  fminbnd, 83
  fspecial, 21, 27, 28
  gcv_tik, 71, 83, 112
  gcv_tsvd, 71, 83, 107
  help, 14
  idcts, 47, 115
  idct2, 34, 48, 75
  idcts2, 34, 47, 48, 116
  ifft2, 34, 41–44, 47, 75
  imadd, 17
  image, 13–16
  imagesc, 2, 3, 13–16, 18, 19
  imdemos, 14
  imdivide, 17
  imfinfo, 13, 14
  imformats, 13, 16
  immultiply, 17
  imnoise, 21, 29
  importdata, 13, 20
  imread, 13, 14, 16, 17, 19, 20
  imshow, 13–16, 18, 19
  imsubtract, 17
  imwrite, 13, 15, 19
  kron, 54
  kronDecomp, 34, 49, 51, 54, 75, 117
  load, 13, 16
  mat2gray, 13, 19
  max, 18
  medfilt2, 28
  min, 18
  norm, 54
  ones, 47
  padPSF, 34, 52, 70, 119
  poissrnd, 29
  psfDefocus, 21, 27, 70
  psfGauss, 21, 27, 52
  rand, 29
  randn, 21, 29, 34, 54

`real`, 43
`rgb2gray`, 13, 17
`save`, 13, 16, 19
`size`, 18
`svd`, 34, 50, 75
`svds`, 34, 49
`tic-toc`, 54
`tik_dct`, 71, 86, 109
`tik_fft`, 71, 86, 108
`tik_sep`, 71, 86, 111
`tsvd_dct`, 71, 86, 104
`tsvd_fft`, 71, 86, 103
`tsvd_sep`, 71, 86, 106
`uint16`, 17, 18
`uint8`, 14, 17, 18
`whos`, 14, 17
`zeros`, 47

matrix
- block circulant with circulant blocks (BCCB), 38, 40–44, 52, 63, 65, 100, 102
- block Hankel with Hankel blocks (BHHB), 38, 40, 44, 46, 52, 58, 66, 101, 102
- block Hankel with Toeplitz blocks (BHTB), 38, 40, 44, 46, 52, 58, 66, 101, 102
- block Toeplitz with Hankel blocks (BTHB), 38, 40, 44, 46, 52, 58, 66, 101, 102
- block Toeplitz with Toeplitz blocks (BTTB), 37, 38, 40, 44, 46, 52, 58, 66, 101, 102
- blurring, 21, 24, 27, 30, 33, 40
- circulant, 34, 36, 39, 40
- condition number, 7, 10, 55, 59
- Hankel, 34, 37, 40
- Kronecker product, 39, 40, 48, 52, 63, 66, 89, 94, 100, 102
- LU factorization, 50
- normal, 33
- orthogonal, 55
- rank-one, 38
- regularized inverse, 78
- Toeplitz, 34, 36, 39, 40
- Toeplitz-plus-Hankel, 37, 39, 40
- trace, 80
- unitary, 33

noise, 1, 5, 8, 12, 23, 28, 29, 60, 61, 69, 72, 79–81, 84–85
- filtered, 78
- Gaussian, 29, 54, 56, 85
- inverted, 6, 9, 10, 55, 61, 78
- Poisson, 29
- quantization, 29, 85
- readout, 29
- uniform, 29
- white, 29, 54, 56, 85

norm
- ∞-norm, 97
- 1-norm, 97
- 2-norm, 54, 91, 92, 94
- Frobenius, 7, 54, 94, 100
- of the residual, 90
- p-norm, 96
- smoothing, 91, 94, 96–97

normal equation, 91

pixel, 1, 2, 13–15, 17, 19, 23–25, 27–29
PNG, 16, 19
point source, 23–25
point spread function (PSF), 23, 24, 27, 34, 35
- atmospheric turbulence, 23, 25
- center, 24, 26, 35, 37, 42, 45, 47
- doubly symmetric, 45, 47, 48, 52, 75
- Gaussian, 26, 45, 52, 56, 58
- out-of-focus, 54
- separable, 38–40, 48, 51, 52, 75

preconditioner, 100–102

regularization, 71, 75
- error, 78, 79
- matrix, 90
- Tikhonov, 75, 83, 90–92, 94
- TSVD, 56, 73, 83

regularization parameter, 71, 76, 78, 84
- choosing, 79, 82
- Tikhonov, 72, 75
- TSVD, 56, 60, 72, 73, 75, 79

regularized solution, 72, 77, 78, 80, 81
residual, 80

singular value decomposition (SVD), 9, 10, 33, 50, 52, 55, 58, 61
singular values, 9, 10, 51, 59
singular vectors, 9–11, 59, 62, 63
soft focus image, 20
solution
 high-frequency components, 10, 57, 60, 67, 78, 92
 low-frequency components, 61, 67, 78
 naïve, 5–8, 10, 11, 43, 44, 48, 50, 51, 54, 55, 59, 65, 66, 74
 oversmoothed, 56, 57, 61, 78
 truncated SVD, 11
 undersmoothed, 56, 57, 60, 78
spectral coordinate system, 72
spectral decomposition, 33, 41, 45, 61, 63, 74, 78, 83
spectral filtering, 55–57, 59
stencil, 93

TIFF, 15–17
Tikhonov regularization, 72–77, 80, 83, 90–92, 94, 95
 using derivative operators, 90, 92–96
 using smoothing norms, 91
total least squares, 100
total variation, 97
truncated SVD (TSVD), 11, 56, 72–78, 80–83

vec, 8

Wiener filter, 96